Multi-Pitch Estimation

Synthesis Lectures on Speech & Audio Processing

Editor
B.H. Juang, *Georgia Tech*

Multi-Pitch Estimation
Mads Græsbøll Christensen and Andreas Jakobsson
2009

Discriminative Learning for Speech Recognition: Theory and Practice
Xiaodong He, Li Deng
2008

Latent Semantic Mapping: Principles & Applications
Jerome R. Bellegarda
2007

Dynamic Speech Models: Theory, Algorithms, and Applications
Li Deng
2006

Articulation and Intelligibility
Jont B. Allen
2005

Multi-Pitch Estimation

Mads Græsbøll Christensen and Andreas Jakobsson

ISBN: 978-3-031-01430-7 paperback
ISBN: 978-3-031-02558-7 ebook

DOI 10.1007/978-3-031-02558-7

A Publication in the Springer series
SYNTHESIS LECTURES ON SPEECH & AUDIO PROCESSING
Lecture #5
Series Editor: B.H. Juang, Georgia Tech

Series ISSN
Synthesis Lectures on Speech & Audio Processing
Print 1932-121X   Electronic 1932-1678

Multi-Pitch Estimation

Mads Græsbøll Christensen
Aalborg University, Denmark

Andreas Jakobsson
Lund University, Sweden

SYNTHESIS LECTURES ON SPEECH & AUDIO PROCESSING #5

ABSTRACT

Periodic signals can be decomposed into sets of sinusoids having frequencies that are integer multiples of a fundamental frequency. The problem of finding such fundamental frequencies from noisy observations is important in many speech and audio applications, where it is commonly referred to as pitch estimation. These applications include analysis, compression, separation, enhancement, automatic transcription and many more. In this book, an introduction to pitch estimation is given and a number of statistical methods for pitch estimation are presented. The basic signal models and associated estimation theoretical bounds are introduced, and the properties of speech and audio signals are discussed and illustrated. The presented methods include both single- and multi-pitch estimators based on statistical approaches, like maximum likelihood and maximum a posteriori methods, filtering methods based on both static and optimal adaptive designs, and subspace methods based on the principles of subspace orthogonality and shift-invariance. The application of these methods to analysis of speech and audio signals is demonstrated using both real and synthetic signals, and their performance is assessed under various conditions and their properties discussed. Finally, the estimators are compared in terms of computational and statistical efficiency, generalizability and robustness.

KEYWORDS

estimation theory, spectral estimation, pitch estimation, pitch detection, fundamental frequency estimation, statistical signal processing, order estimation, model selection, audio processing, audio analysis, music transcription, speech processing, comb filtering, subspace methods, optimal filtering

To Majbritt and Erik
and
to Ylva, Maja, Lina, Erik, and Olle.

Contents

Preface

This book is the result of four years of research that started with the publication of our first paper on pitch estimation back in 2004 [31]. It contains a collection of methods for pitch estimation (and detection), a pitch estimation toolbox if you will, based mostly on our own publications in the area [21, 22, 23, 24, 25, 26, 27, 30, 31, 32, 33, 34, 81, 85] along with supplementary material, based on the work of other authors, that fits within the framework, complements our work, and is representative of methodologies commonly employed in practical pitch estimation algorithms. All the methods presented in this book have in common that they are based on an estimation theoretical approach to pitch estimation and a parametric model of the observed signal. This book does not address all aspects in relation to pitch estimation, but it does contain a collection of tools that can help in obtaining parameters that describe the important characteristics at signal level. It is our hope that the book will provide the reader with a set of useful tools, inspiration for solving problems in relation to pitch estimation, and an improved understanding of the fundamental problems in relation to pitch estimation. In that spirit, MATLAB implementations of some of the estimators that can be found in this book are available online[1].

The book consists of five chapters: Chapter 1 is basically an introduction to pitch estimation, containing the basic signal models, associated bounds on the accuracy of the parameters of such models, and a discussion of the issues that one has to address in constructing and evaluating pitch estimators. Chapter 2 contains a collection of methods for single- and multi-pitch estimation based on statistical models, in particular maximum likelihood and maximum a posteriori estimation. In Chapter 3, we present a number of classical methods based on filtering and some more recent methods based on the principle of optimal filtering. These are loosely based on Capon's classic optimal beamformer design. Finally, subspace-based methods that rely on the principles of subspace orthogonality and shift-invariance are presented in Chapter 4. In the last chapter, Chapter 5, the parameterization of periodic signals is completed by a presentation of a number of methods for finding the amplitudes and phases given a set of frequencies. Each chapter, except the first, ends with first some experimental results and then a discussion of the fundamental properties of the various methodologies and pitch estimators. It is not within the scope of this book to provide very detailed simulations results; rather we will refer to the papers where such results can be found.

A number of people, our friends, colleagues, and present and former students, have contributed to this work in various ways and deserve to be thanked. First of all, we wish to thank all the co-authors on our papers on pitch estimation, namely Søren Holdt Jensen, Søren Vang Andersen, Petre Stoica, Jesper Højvang Jensen, Jesper Kjær Nielsen, Jesper Rindom Jensen, Samuel D. Somasundaram, and Pedro Vera-Candeas. Also, the following people volunteered (more or less) for proof reading the

[1]http://www.morganclaypool.com/page/multi-pitch

manuscript, and we are thankful for their contributions in the form of either countless corrections or suggestions: Jesper Højvang Jensen, Jan Østergaard, Daniele Giacobello, and Jesper Rindom Jensen[2]. Finally, Mads Græsbøll Christensen also wishes to thank Dan P. W. Ellis, Ken Rose, Bob Sturm, Gäel Richard, Betrand David, Roland Badeau, and A. Taylan Cemgil for hosting him on his travels and for many enlightening discussions on speech and audio processing, estimation theory, and music.

Mads Græsbøll Christensen
Dept. of Electronic Systems
Aalborg University, Denmark

Andreas Jakobsson
Dept. of Mathematical Statistics
Lund University, Sweden

February 2009

---

[2]None of the people mentioned here are related, as far as we know, despite several of them sharing the same last name.

Symbols and Notation

x	Scalar
$\mathbf{x}$	Vector
$\mathbf{X}$	Matrix
$(\cdot)_k$	kth source
$(\cdot)^{(i)}$	ith iteration
$\widehat{(\cdot)}$	Estimate
$(\cdot)^T$	Transposition
$(\cdot)^*$	Complex conjugation
$(\cdot)^H$	Hermitian transposition
$p(\cdot)$	Probability density function
$p(\cdot;\cdot)$	Likelihood function
$p(\cdot|\cdot)$	Conditional probability density function
$E\{\cdot\}$	Expectation operator
$E\{\cdot|\cdot\}$	Conditional expectation operator
$\det(\cdot)$	Matrix determinant
$\operatorname{rank}(\cdot)$	Matrix rank
$\dim(\cdot)$	Subspace dimension
$\mathcal{R}(\cdot)$	Range (column space)
$\|\cdot\|_2$	Vector or matrix 2-norm
$\|\cdot\|_F$	Matrix Frobenius norm
$\mathbf{I}$	Identity matrix
$\mathbf{b}_l$	Vector where $[\mathbf{b}_l]_v = 0$ for $v \neq l$ and $[\mathbf{b}_l]_l = 1$
$[\cdot]_{jk}$	(j,k)th entry of matrix
$\lfloor\cdot\rfloor$	Floor function
K	Number of sources
N	Number of samples (i.e., segment length)
M	Covariance matrix size/sub-vector length
$x(n)$	Observed signal
$\mathbf{x}(n)$	Observed signal vector (length M)
$\mathbf{x}$	Observed signal vector with $M = N$
$X(\omega)$	Fourier transform of $x(n)$
$\Phi_x(\omega)$	Power Spectral Density of $x(n)$
$e(n)$	Observation noise
$\mathbf{R}$	Covariance matrix of $x(n)$

$\mathbf{Q}$	Covariance matrix of $e(n)$
$x_k(n)$	kth source of observed signal
L_k	Number of harmonics of kth source
L	Total number of sinusoids $\sum_{k=1}^{K} L_K$
ω_k	Fundamental frequency of kth source
$a_{k,l}$	Complex amplitude of lth harmonic of kth source
$A_{k,l}$	Amplitude of lth harmonic of kth source
$\phi_{k,l}$	Phase of lth harmonic of kth source
$\psi_{k,l}$	Frequency of the lth harmonic of the kth source
ψ_l	Frequency of the l sinusoids (no source assignment)
$e_k(n)$	Observation noise kth source
$\mathbf{R}_k$	Covariance matrix of $x_k(n)$
$\mathbf{Q}_k$	Covariance matrix of $e_k(n)$
$\boldsymbol{\theta}$	Parameters of all sources
$\boldsymbol{\theta}_k$	Parameters of the kth source

Abbreviations

AIC	Akaike Information Criterion
ANLS	Approximate NLS
APES	Amplitude and Phase EStimation method
CCA	Classical Capon Amplitude estimator
CRLB	Cramér-Rao Lower Bound
DFT	Discrete Fourier Transform
EAA	Extended APES Amplitude estimator
ECA	Extended Capon Amplitude estimator
EM	Expectation Maximization
ESPRIT	Estimation of Signal Parameters via Rotational Invariance Techniques
ESTER	ESTimation ERror
EXIP	EXtended Invariance Principle
FFT	Fast Fourier Transform
LS	Least-Squares
MAFI	MAtched FIlterbank
MAP	Maximum A Posteriori
MCA	Multiple constraint Capon Amplitude estimator
MDL	Minimum Description Length
MIDI	Musical Instrument Digital Interface
ML	Maximum Likelihood
MLE	Maximum Likelihood Estimator
MODE	Method Of Direction Estimation
MSE	Mean Square Error
MUSIC	Multiple Signal Classification
NLS	Nonlinear Least-Squares
PDF	Probability Density Function
PSD	Power Spectral Density
PSNR	Pseudo Signal-to-Noise Ratio (usually in dB)
PSTN	Public Switched Telephone Network
RMSE	Root Mean Square (estimation) Error
SAMOS	Subspace-based Automatic Model Order Selection
SNR	Signal-to-Noise Ratio (usually in dB)
WLS	Weighted Least-Squares

CHAPTER 1

Fundamentals

1.1 INTRODUCTION

In his attempt to solve the heat equation for a metal plate in the early 19th century, Jean Baptiste Joseph Fourier inadvertently discovered what we now refer to as Fourier series, and in doing so, he also revolutionized mathematics. A Fourier series is a decomposition of a periodic function into a possibly infinite sum of sinusoids, i.e., simple oscillating functions. These sinusoids all repeat over the same interval, meaning that they have frequencies that are integer multiples of a fundamental frequency. This book is about estimation of this fundamental frequency from noisy observations, a task often referred to as pitch estimation. Let us explore a bit further what is meant by pitch. In the context of music, the American Standard Association defines the term pitch as "that attribute of auditory sensation in terms of which sounds may be ordered on a musical scale" [4], while the American National Standards definition is "that attribute of auditory sensation in terms of which sounds may be ordered on a scale extending from low to high" [3]. The keywords here are auditory and sensation. In the introduction of [129], this leads the authors to conclude that "the word pitch should not be used to refer to a physical attribute of a sound" [129]. We will do exactly the opposite, although not out of spite, and frequently use the word pitch synonymously with fundamental frequency to refer to the physical attribute of signals associated with Fourier series. In fact, we will use it to refer to the fundamental frequency of periodic signals of any origin. We should, perhaps, make use of the less ambiguous term fundamental frequency, but it is a much less elegant term, especially when considering the estimation of the parameters of multiple such periodic waveforms, i.e., multi-pitch estimation. More specifically, we will use the terms pitch and fundamental frequency to refer to the number of repetitions per time interval of a periodic signal and the term pitch period to refer to the time interval between such repetitions. We will also use the term pitch estimation to cover both single- and multi-pitch estimation.

Sometimes the problem of pitch estimation is referred to as pitch detection. Usually, a detection problem is defined as the problem of accepting or rejecting a hypothesis or accepting one from a set of hypotheses, while an estimation problem has to do with finding a certain number or quantity (see, e.g., [89, 90]). Since pitch as per our definition is generally a continuous phenomenon, the problem of finding the pitch is an estimation problem, while the problem of determining whether any pitch is present is a detection problem. One could of course argue that the problem of finding musical notes is a detection problem, since these are discrete, but pitch estimation extends beyond this problem. Certain problems can be seen as either estimation or detection problems, like the problem of determining the number of periodic components in an observed signal or detecting the presence of tones or ensembles of tones in music–sometimes we may not know whether the underlying quantity we are looking for is, in fact, continuous or discrete. But since even musical tones may be modulated due to, e.g., glissando or vibrato, the problem is generally a continuous one, even if the estimates will later be quantized to a discrete musical scale.

The focus of this book is on high-resolution parametric methods for pitch estimation, i.e., methods based on parametric models of the observed signal that are able to come close to the optimal performance. A second trait of the methods considered in this book is that they are all based on rigorous mathematics, relying on such well-established methodologies as statistics, optimal filtering, linear algebra, and convex and sometimes nonlinear optimization. This leads to an understanding of the conditions under which the methods are expected to work and when and why they may fail. Estimators are typically assessed in terms of bias and variance and the determination of these quantities for various estimator are important tasks, since they provide insights into fundamental properties of an estimator. Bias is defined as the difference between the true parameter value and the expected value of the estimate and if this difference is zero for all possible true parameter values, the estimator is said to be unbiased. When we say that an estimator is good, we generally mean that the bias and variance are low. Sometimes estimators are also referred to as being robust. This generally means that the estimator performs well under adverse conditions like a low number of samples, a low signal-to-noise ratio, or that it is insensitive to assumptions being violated, e.g., the model being wrong.

1.2   RELATED WORK

Beyond the methods discussed in this book, there are of course many others (our sincere apologies to everybody whom we have forgotten to mention). We will now briefly discuss some of those. There exists many so-called non-parametric methods based on, for example, the auto-correlation function (e.g., [139]), the cross-correlation function [167], the averaged magnitude difference function [145], the averaged squared difference function, or the cepstrum [1, 123]. All these methods are all based on measuring the similarity of the signal with a delayed version of the same signal in some sense, as is [117]. The difference between these methods lies essentially in terms of how the similarity is measured. They generally suffer from the problem that the solution is not unique even in the ideal case, i.e., there exists multiple lags for which the signal is similar to itself. Another example of a non-parametric method is the so-called harmonic product spectrum of [124] (see also [42]). For an overview of non-parametric methods, we will refer the interested reader to one of the few other books on pitch estimation, namely [75] (see also [76]) and the classical paper [63]; for examples of more recent work on methods of various origins, see [41, 49, 52, 104, 105, 130, 131, 168, 169, 175] and the references therein. In some cases, it may be difficult to classify methods, like those of [92, 140], which are based on various heuristics and ideas. Another class of methods that will not be treated in this book, but yet deserves to be mentioned here, is the class of methods based on models of the human auditory system. Rather than taking their starting point in the properties of the signal, these methods are based on the properties of the human ear and brain. The rationale is that, apparently, the human auditory system has a remarkable ability to estimate multiple simultaneous pitches and separate the various sources from each other, and, perhaps, by mimicking the process, we can design a system that can do the same. For examples of such methods see [93] and the references therein, and for an overview of all things related to pitch perception, see [129] and the more general text books on psycho-acoustics [119, 181]. We would like to stress that the objective of such methods may be quite different from the objective of the methods presented in this book. The former class of methods may aim at modeling the human auditory system, e.g., predicting the outcome of listening experiments for the purpose of better understanding the human auditory system or deriving models that may be applied in various signal processing tasks in, for example, audio coding or hearing aids [37, 38, 171, 172]. The latter class of methods, i.e., the class of methods considered in this book, is concerned with finding the parameters that are most likely to explain the observed signal, and this is generally a different concept altogether than modeling the peculiarities of the human auditory system (see [20] for a discussion of this and [29] for some examples where it makes more sense). Among the statistical parametric methods an important methodology is notably absent, namely the Bayesian approach (aside from the occasional MAP estimator). The reasons for this are quite simple: the topic is big enough to warrant a book by itself and does not lend itself to a short introduction suitable for this book. Besides, other people are more qualified at writing a book on this topic. We do, however, believe that the field is very promising and that many interesting results have already been published, e.g., [15, 16, 40, 61, 62]. It is especially promising in terms of building hierarchies of estimation and detection problems, incorporating prior information and at extracting abstract high-level information. For examples of this see, e.g., [15, 94]. It should be stressed that many of the methods discussed in this book may be combined with the other methods, which may be especially beneficial for those operating on a higher (semantic) levels, like the Bayesian tracking methods that deal with various aspects of pitch estimation, as these are basically complementary to ours.

1.3   SOME APPLICATIONS

There are many applications in which pitch estimation plays an important role. Below, some of these are listed and discussed. While we will here focus on speech and audio applications, the methods considered in this book can be used for any kind of periodic signal, like electrocardiogram (ECG) signals.

Separation A parameterization of a signal into components allows for a natural separation of sources if the signal components have a close relation to the sources. In the case of periodic signals, the models employed throughout this book form such a model. Once the fundamental frequency of all sources has been found, the remaining parameters are easily found and a full parameterization of the individual sources is available, an idea that has been used in, e.g., [18]. Alternatively, one can directly estimate the signals of the individual sources using the pitch [148, 149, 180]. The sources that can be separated using these models include voiced speech and tones of musical instruments. This means that important problems such as the cocktail party problem can be solved, in principle, via source separation based on pitch estimation. This principle is also employed in many computational auditory scene analysis (CASA) schemes (see, e.g., [176]) and is also commonly believed to be a governing principle in the functioning of the human auditory system [129]. The problem of classifying musical instruments is also easy, relatively speaking, once the fundamental frequencies have been found [96].

Enhancement Signal enhancement, or de-noising, deals with removing unwanted noise from an observed signal. This may be beneficial for many reasons. Human listeners will, for example, tend to suffer from fatigue if they have to listen to noisy signals for an extended period of time. Also, the perceived quality of a signal can often be improved by removing the noise. Enhancement algorithms are commonly used as a front-end processor for speech recognizers and coders so that these systems do not have to incorporate models of the background noise in the processing. In this connection, it should be noted that the use of such pre-processing only makes sense if the subsequent processing is suboptimal. Using parametric models, the enhancement problem is almost trivially solved, meaning that it is a matter of finding good estimates of the parameters that describe the signal of interest, whereby the effect of the noise is minimized. In practice, this is done by taking the characteristics of the noise and the properties of the signal of interest into account in the estimation process. An example of a practical speech enhancement systems based on a parametric model can be found in [84].

Compression The signal models used in this book can also form a basis for signal compression. Audio signals are often stationary (i.e., their characteristics stay the same) over about 30 ms and sampled at sampling frequencies of 44.1 kHz, this leads to segments leading to segments consisting of 1323 real-valued samples. If such a signal contains a single tone, it means that all the samples can be represented by, and a signal reconstructed from, a fundamental frequency, an amplitude and a phase for each harmonic. This may be useful for several purposes. First of all, these parameters may be quantized and transmitted, i.e., an audio coder can be based on this principle and several such coders exist, e.g., [78, 134] (see also [19, 28] for an overview of the authors' own contributions within this field). Also for coding of speech signals, where voiced speech segments can be efficiently coded as a single periodic source, such models have been applied [110, 115, 144]. Furthermore, pitch estimators are also often used in linear predictive speech coding for so-called long-term predictors (see, e.g., [95]). But it may also be useful for a different purpose, namely low-complexity signal processing. Because the signal is represented using only few parameters compared to the number of samples, it may lead to much faster algorithms by operating directly on the parameters instead of the samples and then reconstruct the processed signal from the parameters (see also the next item).

Modification It is possible to perform many kinds of signal modification based on parametric models. In fact, sometimes otherwise complicated signal modifications become simple when using an appropriate model. Some classical examples of modifications that become relatively simple for a model consisting of a sum of sinusoids is pitch transposition, also known as pitch shifting or pitch-scale modification, and modification of the duration without altering the pitch, sometimes referred to as time-scale modification [57, 58, 113, 114]. Filtering, of course, also becomes a simple operation asymptotically when dealing with sinusoids, and it is also simple to apply dynamic range compression in the reconstruction of audio signals using a parametric model. There exist also many more esoteric modifications, like voice morphing. In the context of speech coding for packet-based networks, it is also possible to compensate for packet losses using sinusoidal models by interpolation or extrapolation in the compressed domain [110, 144].

Transcription The information required to reproduce a piece of music is represented by musicians using sheet music. Transcription of music into sheet music is today mostly done by hand, i.e., by musicians, and it requires years of training to do this, and it is a very time-consuming and thus a costly affair. Sheet music contains such information as time signatures, chords, scales, individual tones, duration, etc., information that is also the key ingredients of Musical Instrument Digital Interface (MIDI) files. An important problem is thus the identification of individual tones, simultaneous tones in chords and ensembles of tones over time that constitute musical scales. The identification of such tones is the same as pitch estimation and detection. Automatic transcription is, therefore, an important application of the methods considered in this book, and it is sometimes considered the holy grail of pitch estimation. For more on the specifics of automatic transcription of music, we refer the interested reader to [15]. A common way of evaluating automatic transcription is by synthesizing the audio from the transcribed data using MIDI synthesis.

Tuning In music, tuning is the process of adjusting a musical instrument such that tones produced by the instrument obey certain relations, referred to as relative tuning, and are consistent with a reference, like the tone A being 440 Hz (others are sometimes used). Tones that are too high are said to be sharp, while tones that are too low are said to be flat. Depending on the instrument, this may be simple or complicated, and the frequency at which this has to be done may vary a lot, as, for example, the tension of the strings on a guitar may vary with the weather and usage. An obvious application of pitch estimators is for tuning of musical instruments, and as any musician will attest, it is critical that pitch estimators are very accurate for this application.

Classification Since the pitch estimated over time essentially contains all tonal information of a piece of music, it can also be used for identifying songs. Similarly, musical styles, periods and genres can largely be recognized by features derived from pitch, such as musical scales. The features of musical sounds is typically divided into pitch and timbre (aside from loudness and spatial location), where timbre is defined as "that attribute of auditory sensation in terms of which a listener can judge that two sounds similarly presented and having the same loudness and pitch are dissimilar" [4]. Timbre is, therefore, related to the spectral envelope and to the distribution of the amplitudes of the harmonics of a periodic source. Timbre forms the basis of many music information retrieval systems, in particular genre classification applications (see, e.g., [170]), and the parametric representations used in this book may, therefore, directly or indirectly, be used for this.

1.4   SIGNAL MODELS

We will now introduce the fundamentals of the signal model and notation that will be here along with some assumptions that are often used. Throughout the book, we will be working on discrete-time signals, i.e., signals that have been sampled, and we will assume that they have been done so in an appropriate manner. First, we introduce a signal, termed a source, consisting a set of complex sinusoids having frequencies that are integer multiples of a fundamental frequency $\omega_k > 0$, corrupted by an additive noise. The individual sinusoids are referred to as harmonics and the sinusoids are said to be harmonically related. Such a signal can be written for $n = 0, \ldots, N - 1$ as

$$x_k(n) = \sum_{l=1}^{L_k} a_{k,l} e^{j\omega_k l n} + e_k(n) \tag{1.1}$$

$$= s_k(n) + e_k(n), \tag{1.2}$$

where $a_{k,l} = A_{k,l} e^{j\phi_{k,l}}$ is the complex amplitude of the lth harmonic of the source, indexed by k, and $e_k(n)$ is the noise, which is assumed to be zero-mean complex-valued. The complex amplitude is composed of a real, non-zero amplitude $A_l > 0$ and a phase $\phi_{k,l}$. The number of harmonics, L_k, is termed the order of the model and it may be considered either known or unknown. In either case, it is required that $\omega_k L_k < 2\pi$. Here, we have implicitly assumed that the model in (1.1) is valid for $n = 0, \ldots, N - 1$, i.e., that the same parameters and noise characteristics accurately describe the signal over that interval. We will refer to this property as stationarity. Note that the model in (1.1) is sometimes referred to in the literature as the harmonic sinusoidal model. All the unknown real parameters of interest we will organize in a vector defined as

$$\boldsymbol{\theta}_k = \begin{bmatrix} \omega_k & A_{k,1} & \phi_{k,1} & \cdots & A_{k,L_k} & \phi_{k,L_k} \end{bmatrix}^T, \tag{1.3}$$

where $(\cdot)^T$ denotes the transpose. The exact definition of this parameter vector may vary depending on the context and the signal model. Consider, for example, that we may also sometimes be interested in the parameters of the noise $e_k(n)$.

Signals of the form (1.1) are periodic, except for the contribution of the noise, with $s_k(n)$ being the periodic part, which is sometimes also referred to as the deterministic part. In fact, any periodic signal can be decomposed using (1.1), except for some pathological examples that, fortunately, are generally not of interest in engineering applications, and the conditions for convergence of Fourier series are trivially fulfilled when dealing with band-limited signals. We refer to signals of the form in (1.1) as single-pitch signals, since all sinusoids can be described by a single pitch. Examples of such signals are voiced speech, wind instruments like trumpets, ECG signals (except for pregnant women), individual notes of guitars, violins, etc. In Figure 1.1, an example of a periodic signal is depicted, more specifically a speech signal sampled at 8 kHz. It can be seen to consist of a basic

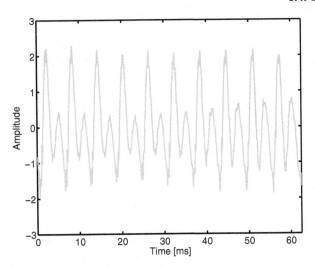

Figure 1.1: Example of a real periodic signal, a voiced speech signal sampled at 8 kHz, having a fundamental frequency of approximately 165 Hz.

repeating waveform. In Figure 1.2, the same signal is shown, but now with Gaussian noise added to it, and its magnitude spectrum can be seen in Figure 1.3. The problem of finding the fundamental frequency of such a signal is referred to as pitch estimation or fundamental frequency estimation. Again, it must be stressed that what we here refer to as pitch is not necessarily the same as *perceived* pitch. Some times the term pitch period is also used in the literature. Here, we will denote this quantity τ_k (in samples). It is related to the fundamental frequency as $\omega_k = 2\pi \frac{1}{\tau_k}$. The fundamental frequency ω_k in radians is related to the physical frequency f_k (in Hertz) as

$$\omega_k = 2\pi \frac{f_k}{f_c}, \qquad (1.4)$$

where f_c is the sampling frequency, which is here the number of complex samples per second, as indicated by subscript c (see Appendix A for an explanation), which is related to the number of real samples per second as $2f_c = f_r$. The signal models that are used in this book are all based on complex representations. The reasons for this are twofold. Firstly, the complex representation leads to a simpler notation and requires less bookkeeping, and, secondly, it leads to computationally more efficient algorithms as only half the number of sinusoidal components is required and only half as many samples are required – when using complex algebra. As speech and audio signals are real, we will have to map the real signals to complex signals; this can be done using the discrete-time analytic signal as described in Appendix A. It should be stressed that for very low fundamental frequencies, it may be necessary to use the real signal model to obtain accurate estimates (see also Section 1.6).

Although the title of this book suggests that we are interested in the fundamental frequency, we will also discuss methods for finding the other signal parameters and in some cases also the noise

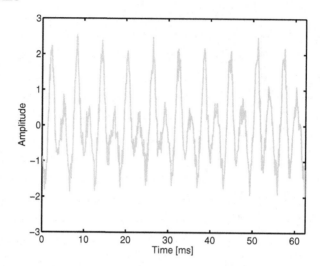

Figure 1.2: Example of a noisy periodic signal, here the signal in Figure 1.1 in additive white Gaussian noise.

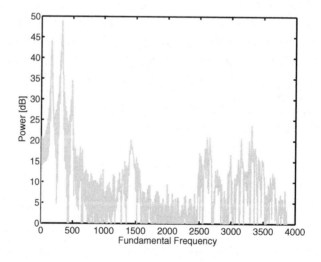

Figure 1.3: Power spectrum of the periodic signal in Figure 1.2.

characteristics. This is useful for many applications, for example when the signal is to be reconstructed using the signal model as is the case in coding applications. The reason for the emphasis on the fundamental frequency is that this problem is much harder than finding the amplitudes and phases. Indeed, for a known fundamental frequency the problem of finding the complex amplitudes is quite simple in comparison, since these parameters are linear.

In many cases, like in most music, the observed signal consists of many such periodic signals, in which case the signal model should be written as

$$x(n) = \sum_{k=1}^{K} x_k(n) = \sum_{k=1}^{K} \sum_{l=1}^{L_k} a_{k,l} e^{j\omega_k ln} + e(n) \tag{1.5}$$

$$= s(n) + e(n). \tag{1.6}$$

We refer to such signals at multi-pitch signals and the model as the multi-pitch model. For this model, the full parameter vector is

$$\boldsymbol{\theta} = \begin{bmatrix} \boldsymbol{\theta}_1^T & \cdots & \boldsymbol{\theta}_K^T \end{bmatrix}^T. \tag{1.7}$$

Chords played on an instrument are examples of signals that can be described by (1.5). In this sense, the term source refers not to individual instruments but to individual notes played on instruments. Recorded speech signals where many speakers speak at the same time are also examples of multi-pitch signals. It is also important to note that even signals that originally are single-pitch may be rendered multi-pitch signals by various processing or the environment. For example, reverberation in a room may lead to the observed signal consisting of several single-pitch signals. The single-pitch case can be described as either a special case of (1.5) with $K = 1$ or as the case where the individual sources in (1.1) are observed. In Figure 1.5, an example of a multi-pitch signal is shown. It contains the sum of the speech signal in 1.1 and the speech signal in Figure 1.4, the latter having a fundamental frequency of approximately 205 Hz. As can be seen, the signal appears to be much more complicated than any of the individual sources.

Sometimes it is useful to disregard the structure of the above signal model at first and consider the following instead:

$$x(n) = \sum_{k=1}^{K} \sum_{l=1}^{L_k} a_{k,l} e^{j\psi_{k,l} n} + e_k(n) \tag{1.8}$$

$$= \sum_{l=1}^{L} a_l e^{j\psi_l n} + e(n), \tag{1.9}$$

where we will use a single subscript when the source assignment is not important and a double subscript for the frequency when it is. Based on this model, one could, for example, estimate the $L = \sum_{k=1}^{K} L_k$ frequencies $\{\psi_l\}_{l=1}^{L}$ and then fit these estimates to form an estimate of ω_k. We will make use of this model from time to time and we will refer to $\{\psi_l\}_{l=1}^{L}$ as the set of unconstrained frequencies and (1.9) as the unconstrained model.

As mentioned, we are here primarily interested in estimating the fundamental frequency. However, it must be emphasized that even when only a subset of the parameters is of interest, the rest may have to be estimated, implicitly, to yield correct estimates. For example, the background noise, which may or may not be produced by the instrument or the speaker, can be white or colored

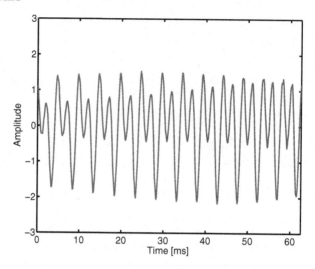

Figure 1.4: Another periodic signal, also a voiced speech signal sampled at 8 kHz, where the fundamental frequency is about 205 Hz.

Gaussian noise, and the number of sinusoidal components can be unknown and variable. These conditions will in many cases affect our ability to estimate the fundamental frequency. The noise characteristics is also important in certain applications, like analysis, classification and compression of music signals.

At this point, it is to elaborate on the meaning of the noise term in our signal models. In estimation theory, the sum of sinusoids is often referred to as deterministic signal components characterized by deterministic but unknown parameters. If we know these parameters, we can predict exactly what the signal will look like. In contrast, the noise is considered to be stochastic and it is, therefore, not possible to accurately predict the observed signal, only its general characteristics. In this terminology, all stochastic signal components are considered noise, even if they are desired parts of the signal. In speech and music signals, stochastic signal components may be the bow noise of a violin, or the unvoiced signal components of a speech signal, and these are very important parts of the signal. However, in relation to estimating the deterministic signal parts, these are still referred to as noise, since they will limit the accuracy at which we can expect to determine the parameters.

Many of the methods considered in this book operate on sub-vectors consisting of M consecutive samples of the observed signal as

$$\mathbf{x}(n) = [\, x(n) \, \cdots \, x(n+M-1) \,]^T, \tag{1.10}$$

where $M \leq N$, and similarly for the sources $x_k(n)$ and the noise $e(n)$ leading to vectors $\mathbf{x}_k(n)$ and $\mathbf{e}(n)$. We remind the reader that these signals were defined for $n = 0, \ldots, N-1$ and note that the constant M may be chosen differently in the book depending on the context. Define a matrix

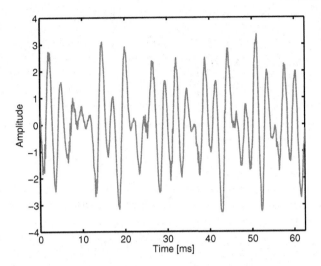

Figure 1.5: An example of a multi-pitch signal consisting the two speech signals, specifically the signals of Figures 1.1 and 1.4. Additionally, white Gaussian noise has been added.

$\mathbf{Z}_k \in \mathbb{C}^{M \times L_k}$ having a Vandermonde structure, being constructed from L_k complex sinusoidal vectors as

$$\mathbf{Z}_k = [\ \mathbf{z}(\omega_k)\ \mathbf{z}(\omega_k 2)\ \cdots\ \mathbf{z}(\omega_k L_k)\ ], \tag{1.11}$$

with $\mathbf{z}(\omega) = [\ 1\ e^{j\omega}\ \cdots\ e^{j\omega(M-1)}\ ]^T$, and a vector containing the complex amplitudes as $\mathbf{a}_k = [\ a_{k,1}\ \cdots\ a_{k,L_k}\ ]^T$. Introducing $z_k = e^{j\omega_k}$, the Vandermonde matrix can be seen to exhibit the following polynomial structure:

$$\mathbf{Z}_k = \begin{bmatrix} 1 & 1 & \cdots & 1 \\ e^{j\omega_k} & e^{j\omega_k 2} & \cdots & e^{j\omega_k L_k} \\ \vdots & \vdots & \ddots & \vdots \\ e^{j\omega_k(M-1)} & e^{j\omega_k(M-1)2} & \cdots & e^{j\omega_k L_k(M-1)} \end{bmatrix} \tag{1.12}$$

$$= \begin{bmatrix} 1 & 1 & \cdots & 1 \\ z_k^1 & z_k^2 & \cdots & z_k^{L_k} \\ \vdots & \vdots & \ddots & \vdots \\ z_k^{(M-1)} & z_k^{(M-1)2} & \cdots & z_k^{(M-1)L_k} \end{bmatrix}. \tag{1.13}$$

Using these definitions, the signal model in (1.5) can be written as

$$\mathbf{x}(n) = \sum_{k=1}^{K} \mathbf{Z}_k \begin{bmatrix} e^{j\omega_k 1n} & & 0 \\ & \ddots & \\ 0 & & e^{j\omega_k L_k n} \end{bmatrix} \mathbf{a}_k + \mathbf{e}(n) \tag{1.14}$$

$$\triangleq \sum_{k=1}^{K} \mathbf{Z}_k \mathbf{a}_k(n) + \mathbf{e}(n) \tag{1.15}$$

or as

$$\mathbf{x}(n) \triangleq \sum_{k=1}^{K} \mathbf{Z}_k(n) \mathbf{a}_k + \mathbf{e}(n). \tag{1.16}$$

As can be seen, either the complex amplitudes or the Vandermonde matrix can be seen as being varying over n, i.e., $\mathbf{a}_k(n) = \mathbf{D}_n \mathbf{a}_k$ and $\mathbf{Z}_k(n) = \mathbf{Z}_k \mathbf{D}_n$ with

$$\mathbf{D}_n = \begin{bmatrix} e^{j\omega_k 1n} & & 0 \\ & \ddots & \\ 0 & & e^{j\omega_k L_k n} \end{bmatrix}. \tag{1.17}$$

Regarding the dimensions of these matrices, we assume throughout the book that the number of samples N and the sub-vector length M are chosen larger than the total number of unknown parameters, i.e., the matrices $\mathbf{Z}_k$ as well as the augmented matrix $\mathbf{Z} = [\ \mathbf{Z}_1 \ \cdots \ \mathbf{Z}_K \ ]$ are assumed to be tall and have full rank for a distinct set of frequencies.

1.5   COVARIANCE MATRIX MODEL

We will make extensive use of the covariance matrix of the sub-vectors. This matrix is defined as

$$\mathbf{R} = \mathrm{E}\left\{\mathbf{x}(n)\mathbf{x}^H(n)\right\}. \tag{1.18}$$

Here, $\mathrm{E}\{\cdot\}$ and $(\cdot)^H$ denote the statistical expectation and the conjugate transpose, respectively. In practice, the covariance matrix is unknown and is generally replaced by the sample covariance matrix defined as

$$\widehat{\mathbf{R}} = \frac{1}{N-M+1} \sum_{n=0}^{N-M} \mathbf{x}(n)\mathbf{x}^H(n). \tag{1.19}$$

Clearly, for $\widehat{\mathbf{R}}$ to be invertible, we require that $M < \frac{N}{2} + 1$. In the following, we will assume that M is chosen accordingly whenever the inverse of the covariance matrix is used. We will also generally let the dimensions of the covariance matrix be proportional to the number of samples as the performance turns out to depend on both parameters (see [157] for an explanation of this phenomenon).

As is well-known, the covariance matrix estimate in (1.19) is the maximum likelihood estimate for Gaussian distributed signals provided that the sub-vectors are independent. Furthermore, it is generally consistent for ergodic processes regardless of their distributions, meaning that the estimate will tend towards the true covariance matrix as the number of observations goes to infinity.

In deriving the estimators, we will make extensive use of the properties of the ideal covariance matrix in (1.18), but in practice it will be replaced by the sample covariance matrix as defined in (1.19). It should be clear from the context which of the two we are referring to. Assuming that the sources are statistically independent, the covariance matrix of the observed signal can be written as

$$\mathbf{R} = \sum_{k=1}^{K} \mathbf{R}_k = \sum_{k=1}^{K} \mathrm{E}\left\{ \mathbf{x}_k(n)\mathbf{x}_k^H(n) \right\}. \tag{1.20}$$

Inserting the signal model for the individual sources in this expression, we can write the covariance matrix as

$$\mathbf{R} = \sum_{k=1}^{K} \mathrm{E}\left\{ (\mathbf{Z}_k\mathbf{a}_k(n) + \mathbf{e}_k(n)) (\mathbf{Z}_k\mathbf{a}_k(n) + \mathbf{e}_k(n))^H \right\} \tag{1.21}$$

$$= \sum_{k=1}^{K} \mathbf{Z}_k \mathrm{E}\left\{ \mathbf{a}_k(n)\mathbf{a}_k^H(n) \right\} \mathbf{Z}_k^H + \mathrm{E}\left\{ \mathbf{e}_k(n)\mathbf{e}_k^H(n) \right\} \tag{1.22}$$

$$= \sum_{k=1}^{K} \mathbf{Z}_k\mathbf{P}_k\mathbf{Z}_k^H + \mathbf{Q}, \tag{1.23}$$

where $\mathbf{Z}_k$ is assumed to be deterministic. The matrix $\mathbf{Q}$ is the covariance matrix of $e(n)$, i.e.,

$$\mathbf{Q} = \mathrm{E}\left\{ \mathbf{e}(n)\mathbf{e}^H(n) \right\} \tag{1.24}$$

$$= \sum_{k=1}^{K} \mathrm{E}\left\{ \mathbf{e}_k(n)\mathbf{e}_k^H(n) \right\} = \sum_{k=1}^{K} \mathbf{Q}_k \tag{1.25}$$

also referred to as the noise covariance matrix with $\mathbf{Q}_k$ denoting the noise covariance matrix of the kth source observation noise, $e_k(n)$. The matrix $\mathbf{P}_k$ is the covariance matrix of the amplitudes, i.e.,

$$\mathbf{P}_k = \mathrm{E}\left\{ \mathbf{a}_k(n)\mathbf{a}_k^H(n) \right\}. \tag{1.26}$$

If the phases of the harmonics are statistically independent and uniformly distributed on the interval $(-\pi, \pi]$, this matrix can be simplified as

$$\mathbf{P}_k = \mathrm{diag}\left(\begin{bmatrix} A_{k,1}^2 & \cdots & A_{k,L_k}^2 \end{bmatrix} \right). \tag{1.27}$$

We will make extensive use of this assumption throughout the book. The more specific assumptions with respect to the observed signal vary depending on the methodology in question. Some methods rely on the observation noise being white and Gaussian distributed, while others rely on it being white. Some methods require that the number of harmonics and sources are known, while, in other cases, these can be estimated jointly with the fundamental frequency, amplitudes and so forth. In this section, we have only mentioned the most general assumptions and we will discuss the more specific ones in the respective chapters.

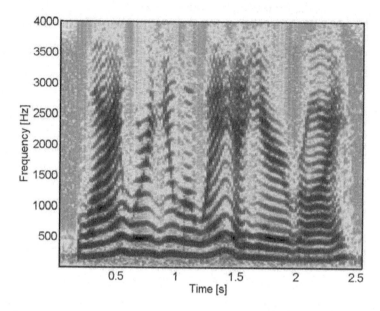

Figure 1.6: Spectrogram of voiced speech signal, utterance: "Why were you away a year, Roy?".

1.6   SPEECH AND AUDIO SIGNALS

Having introduced the fundamental signal model that we will use throughout this book, it is time to discuss whether this model is appropriate for speech and audio signals. Audio signals are particularly troublesome due to the wide variety of such signals. Even when considering only man-made signals such as those produced by musical instruments the variation is overwhelming. It is perhaps impossible to construct a signal model that can explain all phenomena that can be observed in audio signals while also arriving at a useful estimator. In many cases, it is possible to construct dedicated solutions that are based on the specific properties of the observed signals, like, for example, the methods based on physical models of the piano. Such methods may be useful for tuning of pianos and automatic transcription of solo piano music but may be entirely inappropriate for other musical instruments. Below we list and discuss the major problems in relation to estimation theoretical approaches for fundamental frequency estimation and the choices that the algorithm designer has to make.

Non-stationarity Most methods rely on the signal being stationary within the observation time, i.e., in a particular segment, but both speech and music may be subject to changes. The pitch of voiced speech tends to evolve continuously and an extreme example of this are diphthongs. This is illustrated in Figure 1.6, showing the spectrogram of a voiced speech signal. Music on the other hand, is comprised of discrete tones and may thus be subject to discontinuous changes from one tone to another, but it may also be subject to continuous changes in pitch when, for example, vibrato or glissando is applied. These problems may often not be too much of a concern and can be alleviated by simply using short segments, but, depending on the particular kind of scheme, it may be possible to account for non-stationarity in the estimation process by polynomial amplitude and phase models, e.g., [61] or the tracking schemes in [15, 35, 46, 135]. Another possibility is to use an adaptive segmentation of the signal. It is, actually, possible to do this in an optimal way for an additive cost function, like one based on log-likelihoods (see [132] and [133] for some examples of this, though not based on a statistical framework).

Noise Characteristics Estimators are based, implicitly or explicitly, on various assumptions concerning the observed signal and some of these assumptions pertain to the nature of the stochastic signal components also known in estimation theoretical terms as observation noise. There are several aspects of the noise characteristics that may be of concern to us, primarily the noise PDF. It is highly estimator-dependent how critical these assumptions are. It is, for example, well-known that for white Gaussian noise, the least-squares estimator is equivalent to the maximum likelihood estimator and is thus optimal for sufficiently large number of samples, but, for certain problems, it is still asymptotically efficient even if the noise is colored [153]. The subspace methods, on the other hand, are based on the noise being white, but do not rely on it having any specific PDF. It does, however, appear that the assumption of the stochastic signal components being Gaussian but colored is generally sufficient in speech and audio processing, as these assumptions are very common in the literature. Sometimes, the noise is also allowed to have a time-varying variance. In certain problems, the Gaussian assumption seems to be a very poor fit of the observed data. This is, for example, the case when estimating the parameters of an auto-regressive model for voiced speech segments. A Laplacian assumption has been shown to lead to better and more robust estimates [59]. For an interesting read on the history of the Gaussian distribution, its properties, and problems where it appears to be a bad choice, we refer the reader to [91].

Order Estimation Determining the number of sources and the number of harmonics for each source are at the heart of the pitch estimation problem. If these quantities are not known or estimated along the way, the resulting pitch estimates may be non-unique. More specifically, in the context of parametric pitch estimation, the much discussed problems of pitch halvings and doublings[1] are often artifacts of erroneous order estimates. For speech and audio signals, the

---

[1]The focus on these phenomena is a bit curious and sometimes distracting as the problem is more general than that and often only a symptom of a more fundamental issue. In fact, spurious estimates may be obtained for integer ratios, i.e., $\frac{q}{p}$ with $q, p \in \mathcal{N}$, of the true fundamental frequency.

model order may vary from one segment to another and it is, therefore, necessary to estimate it on a frame-by-frame basis. This phenomenon can clearly be seen from the spectrogram in Figure 1.6. It is also worth noting that the Cramér-Rao lower bound for the fundamental frequency estimation problem indicates that it is actually beneficial to estimate the order adaptively over time to achieve the highest possible accuracy (see Section 1.8). The number of harmonics that can be said to be in the signal with any statistical certainty will also depend on the SNR as harmonics having small amplitudes may be completely buried in noise.

Model Selection The problem of model selection is intimately related to the model order estimation problem. As we have discussed, different models may apply to different kinds of signals. An estimator that should be valid for a large range of instruments and signals should thus take this into account. One way to do this is to test different models for a particular signal and then use the one that is most likely to explain the observed signal. For example, speech signals are often classified as either being periodic (voiced) or a realization of a stochastic process, e.g., a colored Gaussian process (unvoiced). A different approach is to use a very general model that can account for all (or most) phenomenon from which it is possible to extract simpler structures. An example of such a model is the perturbed model of [34], which reduces to the usual harmonic model when the perturbations are small (see also Section 1.7 for more on this).

Overlapping Harmonics The next issue is that of spectrally overlapping harmonics, meaning the problem of interaction between either the harmonics of one source or the harmonics of different sources. This happens when the frequencies of two sinusoids are close to each other relative to the number of observations. This can of course always happen for a multi-pitch signal, but it may happen also for any single-pitch signal when the fundamental frequency is very low and/or the number of samples is low. The problem is twofold. Firstly, a low number of samples will cause spectral overlap, i.e., non-zero inner products, of the harmonics for a low fundamental frequency. Secondly, it will cause spectral overlap between the low harmonics and their complex conjugates. For some of the estimators considered in this book, it is straightforward, but computationally demanding, to take this phenomenon into account, like for the least-squares based methods, while it is more difficult for others. It should be noted that the accuracy at which it is mathematically possible to estimate the fundamental frequency will depend on such phenomena, and, as the fundamental frequency tends towards zero, it will, in fact, become mathematically impossible to estimate the pitch [23]. Interaction between harmonics is a fairly predictable phenomenon, though, for the single-pitch case, but, for multi-pitch signals, it can become very complicated. This is especially true for music signals, since musical chords are comprised of sources whose fundamental frequencies are related by simple relations, like a root 5 chord where the second note of the chord has a fundamental frequency, which is approximately 1.5 of the root. It is easy to imagine that this can cause all sorts of problems.

Inharmonicity The harmonic signal is based on the assumption that the harmonics have frequencies that are integer multiples of the fundamental. This assumption does not always hold, for instance, as the stiffness of a string of a musical instrument may cause the harmonics to be stretched meaning that they have frequencies that are higher than the integer multiples and get progressively further from it for higher harmonics [53, 54]. This phenomenon is known as inharmonicity and can be observed from the signal shown in Figure 1.7 as it is not perfectly periodic. Both the problem of inharmonicity and non-stationarity can also be seen as problems related to model mismatch, meaning that if the integer relationship between the fundamental and its harmonics is used, the model we use does not fit the observed signal. Model mismatch often leads to biased parameter estimates and this is exactly what happens if the inharmonicity phenomenon is ignored for, e.g., a piano [34] (see also [92]).

Missing Harmonics It sometimes happens that a signal consists of only a subset of the integer multiples of the fundamental. For example, in the PSTN telephone network, speech signals are commonly high-pass filtered at about 300 Hz, meaning that the lower harmonics for speakers having pitches below 300 Hz will be missing from the observed signal. Similarly, a signal may consist of only even or odd harmonics, or some harmonics may be so buried in noise that it is not possible to determine whether they are there or not. For some methods, such phenomena may not cause any problems as a zero amplitude sinusoid does not make any difference and the fundamental frequency capturing the frequencies of the harmonics is still the same, while for others, like subspace methods, it may be critical to account for. In some cases, like for the PSTN telephone network, we know that the harmonics are missing and can easily take the high-pass filter into account. There may be cases, though, where we cannot know this. Generally, these problems can be alleviated by considering different patterns, like only odd or even harmonics, and then seeing the problem as a joint pitch estimation and model selection problem. For the methods considered in this book that can take different model orders into account, it is usually also straightforward to extend those principles to account for missing harmonics, but it will usually lead to an unnecessarily complicated notation, which is why we will not discuss the problem any further.

It is our opinion that estimation theory is the art of constructing simple and elegant methods that solve real-life problems based on rigorous mathematics, be it statistics or linear algebra. To arrive at such methods, it may be necessary to make a number of simplifying assumptions. But by being explicit about these assumptions, an understanding of the properties of the method can be achieved and it can be predicted, with high accuracy, when the method will work and when it most likely will not and why. As an example of why simplifications are necessary, consider the multi-pitch estimation problem. It is possible, in principle, to evaluate the likelihood function for all combinations of fundamental frequencies, number of sources, number of harmonics, different noise PDFs and so forth, but it is not simple and generally not useful in practice to do so; the necessary computation time will be prohibitive for most real-life applications.

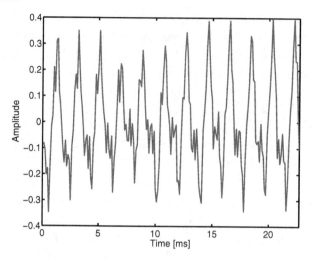

Figure 1.7: A segment of a piano note illustrating the inharmonicity phenomenon.

1.7   OTHER SIGNAL MODELS

It has been observed by many researchers that speech and audio signals are not perfectly harmonic in the real world [53, 58, 103] and this may be important in some applications, even if it does lead to a reasonable fundamental frequency estimate. For example, if a full model of the individual sources is desired, then also the amplitudes have to be estimated. It is easy to imagine that the amplitudes may be wrong if the peak is not located exactly at an integer times the fundamental frequency. For audio coding applications, such errors may be audible and have to be taken into account. A number of phenomena may cause the signals to deviate from perfect periodicity. A well-known example of this is stiff-stringed instruments that give rise to harmonics whose frequencies are not integer multiples of a fundamental. An example of such a model for stiff-stringed instruments is

$$\psi_{k,l} = \omega_k l \sqrt{1 + B_k l^2},\tag{1.28}$$

where $B_k \ll 1$ is an unknown, positive stiffness parameter that varies with the type of string being used. This model is derived by solving differential equations for waves travelling along a string [146]. This model has been used in various methods, e.g., [34, 49, 62].

It is important to note that one can arrive at other models depending on the assumptions one make, with some examples being given in [146]. Also, interaction effects between the bridge of a guitar and its body occur, and these may cause the harmonics to have frequencies that are lower than integer multiples of the fundamental. Interestingly, our terminology of k denoting the source is consistent with this phenomenon as opposed to, e.g., k denoting a particular instrument. It is usually straightforward to incorporate such models in estimators, only one has to either somehow find B_k a priori or estimate it jointly with the other model parameters. The main problem with these models is that they are instrument dependent and sometimes also depend on other things too and one may have to consider many such models to construct a fundamental frequency estimator that may be applicable to a wide range of instruments. An alternative model is the so-called perturbed model where the frequency of each individual harmonic may be shifted by a small perturbation $\Delta_{k,l}$, i.e.,

$$\psi_{k,l} = \omega_k l + \Delta_{k,l}. \tag{1.29}$$

One may reasonably wonder how the introduction of L_k additional nonlinear parameters is supposed to help solve our problem. First of all, this model covers, in principle, all other variations of the harmonic models and the unconstrained model too, if one allows arbitrarily large perturbations. Secondly, it is often possible to solve for the perturbations in a computationally efficient manner, depending on the specific estimator under consideration. And in the sense of the set of perturbations $\Delta_{k,l}$ being stochastic, one can argue that an estimator should be robust to such phenomena. This is the philosophy we will employ in the remainder of the book. It must be emphasized, though, that it is possible to incorporate models like the one in (1.28) in all the estimators considered in this book.

There exists also a number of models and methods that take into account the non-stationary nature of many speech and audio signals, e.g., [61], by introducing, for example, polynomial models of the amplitude and phase. Some of these models can be incorporated in some of the estimators considered in this book, although the use of more complicated signal models often lead to computationally less efficient algorithms and are, therefore, best avoided. It is our opinion, and also our experience, that the problem of non-stationarity is less important than the more fundamental problems of finding the parameters, including the model order and the noise characteristics, that characterize (1.5). More specifically, it is a matter of using short absolute segment lengths so that the signal model is a good approximation of the observed signal. It may of course be the case that the variations of the parameters are of interest as would be the case in transcription of music where bends, slides and vibrato play an important role. In that case, the signal models should be modified accordingly.

1.8   PARAMETER ESTIMATION BOUNDS

It is possible to derive various bounds on the accuracy at which it is possible to estimate some parameters of a model. This is useful for evaluation purposes, but such bounds can also be quite informative, which turns out to be the case for the fundamental frequency estimation problem. When simple closed-form expressions for these bounds exist, one can determine how the problem depends on various conditions, like the number of observations or the signal-to-noise ratio (SNR). The Cramér-Rao lower bound (CRLB) is one such bound for unbiased parameter estimates and we will now introduce the fundamental results related to this bound. For a comprehensive introduction to the CRLB, we refer the interested reader to the excellent book [89] and for examples of other bounds, see [116, 141]. An estimator is said to be unbiased if it results in an estimate $\hat{\theta}_i$ of the ith parameter θ_i of the parameter vector $\boldsymbol{\theta} \in \mathbb{R}^P$ whose expected value is the true parameter, i.e.,

$$E\left\{\hat{\theta}_i\right\} = \theta_i \ \forall \theta_i, \tag{1.30}$$

and the difference $\theta_i - E\left\{\hat{\theta}_i\right\}$, if any, is referred to as the bias. The CRLB of the parameter is given by

$$\mathrm{var}(\hat{\theta}_i) \geq \left[\mathbf{I}^{-1}(\boldsymbol{\theta})\right]_{ii}, \tag{1.31}$$

where $\mathrm{var}(\cdot)$ denotes the variance defined as

$$\mathrm{var}(\hat{\theta}_i) = E\left\{\left(\hat{\theta}_i - E\left\{\hat{\theta}_i\right\}\right)^2\right\}, \tag{1.32}$$

and $\mathbf{I}(\boldsymbol{\theta})$ is the $P \times P$ Fischer information matrix evaluated for the true parameter vector $\boldsymbol{\theta}$, which is here considered to be deterministic but unknown. The Fischer information matrix is defined as

$$\left[\mathbf{I}(\boldsymbol{\theta})\right]_{il} = -E\left\{\frac{\partial^2 \ln p(\mathbf{x}; \boldsymbol{\theta})}{\partial \theta_i \partial \theta_l}\right\}, \tag{1.33}$$

with $\ln p(\mathbf{x}; \boldsymbol{\theta})$ being the log-likelihood function of the observed signal $\mathbf{x} \in \mathbb{C}^N$. The bound in (1.31) holds under the so-called regularity condition (see [89]), which is

$$E\left\{\frac{\partial \ln p(\mathbf{x}; \boldsymbol{\theta})}{\partial \boldsymbol{\theta}}\right\} = \mathbf{0} \quad \forall \boldsymbol{\theta}. \tag{1.34}$$

Here, $\partial \ln p(\mathbf{x}; \boldsymbol{\theta})/\partial \boldsymbol{\theta}$ is defined to be a column vector. In both of the equations above, the expectation is taken with respect to $p(\mathbf{x}; \boldsymbol{\theta})$. For the case where the observation vector is complex Gaussian with mean $\boldsymbol{\mu}(\boldsymbol{\theta})$ and covariance matrix $\mathbf{Q}(\boldsymbol{\theta})$, i.e., $\mathbf{x} \sim \mathcal{N}(\boldsymbol{\mu}(\boldsymbol{\theta}), \mathbf{Q}(\boldsymbol{\theta}))$, the CRLB is given by [153]

$$\begin{aligned}\left[\mathbf{I}(\boldsymbol{\theta})\right]_{il} = \ &\mathrm{Tr}\left\{\mathbf{Q}^{-1}(\boldsymbol{\theta})\frac{\partial \mathbf{Q}(\boldsymbol{\theta})}{\partial \theta_i}\mathbf{Q}^{-1}(\boldsymbol{\theta})\frac{\partial \mathbf{Q}(\boldsymbol{\theta})}{\partial \theta_l}\right\} \\ &+ 2\,\mathrm{Re}\left\{\frac{\partial \boldsymbol{\mu}^H(\boldsymbol{\theta})}{\partial \theta_i}\mathbf{Q}^{-1}(\boldsymbol{\theta})\frac{\partial \boldsymbol{\mu}(\boldsymbol{\theta})}{\partial \theta_l}\right\}.\end{aligned} \tag{1.35}$$

Here, $\partial \boldsymbol{\mu}(\boldsymbol{\theta})/\partial \theta_i$ is defined to be a column vector containing the entries of the vector $\boldsymbol{\mu}(\boldsymbol{\theta})$ differentiated with respect to the ith parameter of $\boldsymbol{\theta}$ and similarly for the matrix derivative.

For the case where the covariance matrix $\mathbf{Q}(\boldsymbol{\theta})$ does not depend on any of the parameters in $\boldsymbol{\theta}$, we denote it $\mathbf{Q}$. Furthermore, under the assumed conditions, (1.35) reduces to

$$\left[\mathbf{I}(\boldsymbol{\theta})\right]_{il} = 2\,\mathrm{Re}\left\{\frac{\partial \boldsymbol{\mu}^H(\boldsymbol{\theta})}{\partial \theta_i}\mathbf{Q}^{-1}\frac{\partial \boldsymbol{\mu}(\boldsymbol{\theta})}{\partial \theta_l}\right\}. \tag{1.36}$$

We will now briefly present the CRLBs associated with finding the parameters of a periodic signal from $x_k(n)$. For simplicity, we consider the case where $e_k(n)$ is white Gaussian and independent having variance σ_k^2. In this case, we have that $\mathbf{Q}_k = \sigma_k^2\mathbf{I}$. Despite these seemingly restrictive assumptions, we can still learn some interesting facts about our problem. For a single source and a high number of samples, i.e., $N \gg 1$, the lower bound on the variance of unbiased parameter estimates can be found using (1.36) and are given by (see [23] and [122] for details)

$$\text{var}(\hat{\omega}_k) \geq \frac{6\sigma_k^2}{N(N^2 - 1)\sum_{l=1}^{L_k} A_{k,l}^2 l^2} \qquad (1.37)$$

$$\text{var}(\hat{A}_{k,l}) \geq \frac{\sigma_k^2}{2N} \qquad (1.38)$$

$$\text{var}(\hat{\phi}_{k,l}) \geq \frac{\sigma_k^2}{2N}\left(\frac{1}{A_{k,l}^2} + \frac{3l^2(N-1)^2}{\sum_{m=1}^{L_k} A_{k,m}m^2(N^2 - 1)}\right). \qquad (1.39)$$

The right-hand side of these equations are referred to as the CRLBs of the respective parameters. The CRLB for the fundamental frequency can be seen to depend on the pseudo signal-to-noise ratio (PSNR), defined in dB as

$$PSNR = 10\log_{10}\frac{\sum_{l=1}^{L} A_{k,l}^2 l^2}{\sigma_k^2} \text{ [dB]}, \qquad (1.40)$$

while the bound for the phase depends on both this quantity and the local SNR, $A_{k,l}^2/\sigma_k^2$. Under the assumption that the sources are independent and that the harmonic frequencies are distinct, (1.37) can also be expected to hold approximately for the problem of estimating the fundamental frequencies in (1.5) for the multi-pitch case, only σ_k has to be replaced by σ. The bounds will not be valid, though, when there is a significant spectral overlap of the harmonics, i.e., for low fundamental frequencies (see [23]) and when the harmonics of a multi-pitch signal overlap. In these cases, the bounds will be much higher, which will be revealed by the exact CRLB [107]. Also, it must be stressed that for a low number of samples, the exact CRLB for the various parameters will depend on the other parameters, including those of other sources.

Interestingly, the CRLBs associated with estimating the parameters of the unconstrained model from $x_k(n)$ are somewhat different. More specifically, these are [88]

$$\text{var}(\hat{\psi}_{k,l}) \geq \frac{6\sigma_k^2}{N(N^2 - 1)A_{k,l}^2} \qquad (1.41)$$

$$\text{var}(\hat{A}_{k,l}) \geq \frac{\sigma_k^2}{2N} \qquad (1.42)$$

$$\text{var}(\hat{\phi}_{k,l}) \geq \frac{2\sigma_k^2(2N - 1)}{A_{k,l}^2 N(N + 1)}. \qquad (1.43)$$

As can be seen, the bounds are different for the frequency parameters and the phases with these depending on the local SNR, while the amplitude bound is the same. This underlines that although the problem of fundamental frequency estimation is related to frequency estimation, the problems and their properties and pitfalls are different. The bounds associated with the fundamental frequency estimation problem suggest that this problem is simpler in the sense that we should be able to do this more accurately than estimating any individual frequencies. Having pondered this, one may suspect that it is a more robust procedure to search directly for the fundamental frequency, and indeed this turns out to be the case. We will return to this later on.

1.9 EVALUATION OF PITCH ESTIMATORS

So how does one go about evaluating pitch estimators? There are several aspects of pitch estimators that need to be evaluated to determine their properties and robustness. Below, some common test methodologies and signals are listed and discussed.

Monte Carlo The Monte Carlo method is a method for random sampling, where the same experiment is repeated many times and each time, all stochastic components are randomized. The stochastic components could, for example, be the noise that is added to a signal or even model parameters, such as the model order, fundamental frequency, amplitude or phase of a periodic source. For each set of realization, parameters of interest are estimated and the statistics of the estimates, like variance and bias, are calculated. The higher the number of times the experiment is repeated, the better, but for complicated nonlinear problems, like the pitch estimation problem, it is often only possible to do this in the order of hundreds of times. The Monte Carlo method can be used in combination with either real or synthetic signals, but are most often used in combination with synthetic signals.

Synthetic Signals Synthetic signals generated using the assumed signal model can be used for answering the following questions: given that the signal model is correct, what is then the accuracy at which we can determine the various parameters, and how does the performance depend on various conditions and user parameters such as the segment length, the signal-to-noise ratio, the sub-vector length, and other quantities. This is also a good way to ensure that the estimators have been correctly implemented. One can experimentally determine how close the estimator is to the optimal performance by estimating the mean squared estimation error using the Monte Carlo method and compare it to the CRLB and the performance of other estimators. More importantly, estimators frequently exhibit so-called thresholding behavior for nonlinear estimation problems, meaning that below a certain number of samples (or PSNR), the method ceases to work and results in non-informative estimates. Monte Carlo simulations using synthetic signals can help to determine under which circumstances this will happen. Using synthetic signals, it can also be determined how sensitive the method is to various assumptions by violating these assumptions in simulations, for example, by using colored noise instead of white noise.

MIDI Signals Signals generated using MIDI are, perhaps, best characterized as being somewhere in-between synthetic and real signals, depending on the type and quality of the synthesizer used.

The good thing about using MIDI generated signals is that the ground truth data (i.e., the true pitch) is readily available and that the environment is fairly controllable and the experiments reproducible. But it is not always simple to determine exactly what kind of synthesis is used and/or how close to real signals the generated signals are; some may be based on real sampled signals, while others may be based on physical models of musical instruments or extremely simplistic mathematical tricks. One can, though, safely assume that they are generated using a model that is at least as complicated as those of equations (1.1) and (1.5). To be safe, one can of course use several different MIDI synthesizers.

Real Signals The synthetic signal methods do not answer the very important question of whether the assumed signal model is in fact correct. Only the application of the algorithm to real signals can answer that. There are many databases of real music and speech signals available on the Internet, but there are two major problems with using real data: a) often no ground truth data is available, meaning that we have nothing but our own expectations to compare the estimates with, which makes the results rather subjective, and b) even when such ground truth data is available, it is usually not available at the accuracy needed to determine whether an estimator is optimal or how far it is from being optimal, and usually only crude performance measures can be applied in these types of evaluations. Furthermore, even when some kind of ground truth data is available, it may be of course be completely wrong, or the pitch may be ambiguous. Consider, for example, a segment containing silence and then the onset of a tone. Is there a pitch in such a segment or is an algorithm right in saying that there is only silence? Or what about the more complicated creaky voice phenomenon (see [99])? It is often the case that researchers have resorted to pruning such cases manually, as is the case in the results reported in [41]. This, of course, introduces an arbitrary and subjective element in the evaluation and it is then meaningless to compare results at a high granularity. In many cases, though, real signals can be used for providing illustrative examples, which can be used for qualitative analysis of the behavior of estimators. For example, by analyzing the examples where a particular method appears to fail, care should, however, be taken in choosing the signals. Some signals exhibit higher degrees of stationarity and inharmonicity than others. The Monte Carlo method can be applied to real signals by adding, for example, noise to the signal to analyze the behavior of the estimates as a function of the SNR or other parameters.

It is often recommendable to keep the experiments as simple as possible, at least at first; we cannot possibly hope to solve complicated problems if we cannot even solve simple ones. The authors of this book recommend that the following tests be performed when evaluating the performance of pitch estimators: most importantly, the variance and bias of the estimator should be evaluated under various conditions, like different N, M, ω_k, L_k, K, PSNR, SNR, and different amplitude distributions using the Monte Carlo method and synthetic signals generated using the assumed signal model. A good measure is the root mean square estimation error (RMSE), which is defined as

$$RMSE = \sqrt{\frac{1}{SK} \sum_{k=1}^{K} \sum_{s=1}^{S} \left(\hat{\omega}_k^{(s)} - \omega_k \right)^2},$$ (1.44)

where ω_k and $\hat{\omega}_k^{(s)}$ are the true fundamental frequency and the estimate for source k in Monte Carlo iteration s, and with S being the number of Monte Carlo trials. By comparing the RMSE to the CRLB, this quantity determines a) how close the method is to being optimal under good conditions, and b) the thresholding behavior of the estimators, i.e., under which conditions the estimator can be expected to work and how it compares to other estimators. The robustness of the estimator towards the issues discussed in Section 1.6 can also be evaluated using the Monte Carlo method based on synthetic signals. In doing so, it is likely that several assumptions will be violated, and this will reveal the robustness of the estimator towards such violations. The key point here is that these are likely to occur when dealing with real speech and audio signals.

CHAPTER 2

Statistical Methods

2.1 INTRODUCTION

The basic idea of statistical estimation methods is to pose a model of the observed signal in terms of a probability density function (PDF), which is parameterized by in our case the parameters of sets of harmonically related sinusoids. Within this framework, the parameters of these sinusoids are often viewed as being deterministic but unknown, while the observation noise, i.e., any stochastic signal components, are considered random. These deterministic but unknown parameters can then be estimated by maximizing the so-called likelihood of the observed signal, i.e., the parameters that are most likely to explain the observed signal are the maximum likelihood estimates. The objective function is referred to as the likelihood function as it is viewed as a function of the unknown parameters rather than the observed signal. The principle of maximum likelihood estimation is probably the most commonly used and estimators based on it are well-known to have excellent performance for a large number of samples. For certain cases, like the Gaussian case, the maximum likelihood estimator also has a simple form, in fact, it is the minimizer of a weighted squared error and for white Gaussian observation noise, it is the even simpler minimizer of the 2-norm, i.e., the least-squares estimator. It is, however, complicated to derive computationally efficient methods for finding nonlinear parameters like the frequencies of sinusoids and this is exactly the goal of this chapter. We will first present the fundamentals of maximum likelihood estimation and maximum a posteriori (MAP) estimation, and we then show how these principles can be used in various ways for finding the fundamental frequency of periodic signals and solve some of the associated problem such as model selection and order estimation.

2.2 MAXIMUM LIKELIHOOD ESTIMATION

It seems natural to adapt this well-merited scheme to our problem and we will here first consider only the single-pitch case, i.e., $K = 1$. The method can be generalized to the multi-pitch scenario, but in its exact form, it is not very useful, since it leads to a complicated multi-dimensional nonlinear optimization problem, but later we will consider an approximate solution. We will now present the maximum likelihood method for finding the parameters of a single periodic signal. The algorithm operates on a signal sub-vector at time n, defined as

$$\mathbf{x}_k(n) = [\, x_k(n) \; \cdots \; x_k(n + M - 1) \,]^T , \tag{2.1}$$

which is constructed from the observed single-pitch signal $x_k(n)$. For many signals, such a sub-vector can be modeled as a sum of L_k harmonically related complex sinusoids in colored Gaussian noise

$\mathbf{e}_k$ having covariance matrix $\mathbf{Q}_k$, i.e.,

$$\mathbf{x}_k(n) = \mathbf{s}_k(n) + \mathbf{e}_k(n) \tag{2.2}$$
$$= \mathbf{Z}_k(n)\mathbf{a}_k + \mathbf{e}_k(n), \tag{2.3}$$

with $\mathbf{a}_k = [\ A_{k,1}e^{j\phi_{k,1}}\ \cdots\ A_{k,L_k}e^{j\phi_{k,L_k}}\ ]^T$ being a vector containing the complex amplitudes. Furthermore, $\mathbf{Z}_k(n)$ is a Vandermonde matrix at time n, defined as

$$\mathbf{Z}_k(n) = [\ \mathbf{z}_1(n)\quad\cdots\quad\mathbf{z}_{L_k}(n)\ ], \tag{2.4}$$

where the mth entry of the column vector $\mathbf{z}_l(n) \in \mathbb{C}^M$ is defined as $[\mathbf{z}_l(n)]_m = e^{j\omega_0 l(n+m-1)}$. Since $\mathbf{x}_k(n)$ has length M and we have N observations of $x_k(n)$, we can thus construct a set of $G = N - M + 1$ different sub-vectors $\{\mathbf{x}_k(n)\}_{n=0}^{G-1}$.

Next, we introduce the signal and noise parameter vector $\boldsymbol{\theta}_k$ containing the fundamental frequency ω_k, the complex amplitudes $\{A_{k,l}e^{j\phi_{k,l}}\}$ and thereby implicitly the order L_k and the noise covariance matrix $\mathbf{Q}_k$ of the model in (2.3). Assuming that $\mathbf{Q}_k$ is invertible, the likelihood function of the observed signal sub-vector $\mathbf{x}_k(n)$ can then be written as

$$p(\mathbf{x}_k(n); \boldsymbol{\theta}_k) = \frac{1}{\pi^M \det(\mathbf{Q}_k)} e^{-\mathbf{e}_k^H(n)\mathbf{Q}_k^{-1}\mathbf{e}_k(n)}, \tag{2.5}$$

with $\det(\cdot)$ denoting the matrix determinant. Now, assuming that the deterministic part $\mathbf{s}_k(n)$ is stationary and $\mathbf{e}_k(n)$ is independent and identically distributed over n, the likelihood of the observed set of vectors $\{\mathbf{x}_k(n)\}_{n=0}^{G-1}$ can be written as

$$\begin{aligned}
p(\{\mathbf{x}_k(n)\}; \boldsymbol{\theta}_k) &= \prod_{n=0}^{G-1} p(\mathbf{x}_k(n); \boldsymbol{\theta}_k) \\
&= \frac{1}{\pi^{MG} \det(\mathbf{Q}_k)^G} e^{-\sum_{n=0}^{G-1} \mathbf{e}_k^H(n)\mathbf{Q}_k^{-1}\mathbf{e}_k(n)}.
\end{aligned} \tag{2.6}$$

Although the approach of splitting the signal into sub-vectors $\mathbf{x}(n)$ is inherently suboptimal, since it ignores inter-vector dependencies, it is required in order to estimate the signal and noise covariance matrices. Taking the logarithm of (2.6), we get the so-called log-likelihood function, i.e.,

$$\begin{aligned}
\mathcal{L}(\boldsymbol{\theta}_k) &= \sum_{n=0}^{G-1} \ln p(\mathbf{x}_k(n); \boldsymbol{\theta}_k) \\
&= -GM \ln \pi - G \ln \det(\mathbf{Q}_k) - \sum_{n=0}^{G-1} \mathbf{e}_k^H(n)\mathbf{Q}_k^{-1}\mathbf{e}_k(n).
\end{aligned} \tag{2.7}$$

The maximum likelihood estimates of the parameters $\boldsymbol{\theta}_k$ are then

$$\hat{\boldsymbol{\theta}}_k = \arg\max \mathcal{L}(\boldsymbol{\theta}_k), \tag{2.8}$$

which are found by estimating the noise vector as

$$\hat{\mathbf{e}}_k(n) = \mathbf{x}_k(n) - \hat{\mathbf{s}}_k(n) \tag{2.9}$$

where the deterministic part of the signal model $\hat{\mathbf{s}}_k(n)$ is constructed from the parameter estimates.

The main difficulty in evaluating this cost function is that the fundamental frequency is a nonlinear parameter and that the noise covariance matrix $\mathbf{Q}_k$ generally is unknown. The amplitudes and phases, on the other hand, can be seen in (2.3) to be linear complex parameters that can easily be found given the fundamental frequency and the noise covariance matrix. Given the (invertible) noise covariance matrix $\mathbf{Q}_k$, the complex amplitudes $\mathbf{a}_k$ can be found, for a certain candidate fundamental frequency, using weighted least-squares (WLS) as

$$\hat{\mathbf{a}}_k = \left(\sum_{n=0}^{G-1} \mathbf{Z}_k^H(n) \mathbf{Q}_k^{-1} \mathbf{Z}_k(n) \right)^{-1} \sum_{n=0}^{G-1} \mathbf{Z}_k^H(n) \mathbf{Q}_k^{-1} \mathbf{x}(n), \tag{2.10}$$

where the inverse exists for tall or square $\mathbf{Z}_k$, i.e., for $L_k \leq M$, constructed from a distinct set of frequencies.

2.3   NOISE COVARIANCE MATRIX ESTIMATION

Next, we will address how to estimate the noise covariance matrix in an efficient manner. Defining $\mathbf{Z}_k = \mathbf{Z}_k(0)$ and assuming that the phases of the harmonics are independent and uniformly distributed on the interval $(-\pi, \pi]$, the covariance matrix $\mathbf{R}_k \in \mathbb{C}^{M \times M}$ of the signal in (2.3) can be written as (see [155, 156]),

$$\mathbf{R}_k = \mathrm{E}\left\{ \mathbf{x}_k(n) \mathbf{x}_k^H(n) \right\} = \mathbf{Z}_k \mathbf{P}_k \mathbf{Z}_k^H + \mathbf{Q}_k \tag{2.11}$$

where

$$\mathbf{P}_k = \mathrm{E}\left\{ \mathbf{a}_k \mathbf{a}_k^H \right\} = \mathrm{diag}\left(\begin{bmatrix} A_{k,1}^2 & \cdots & A_{k,L_k}^2 \end{bmatrix} \right). \tag{2.12}$$

The noise covariance matrix $\mathbf{Q}_k$ is most often unknown and an estimate has to be obtained. Here, we will do this based on the signal covariance model in (2.11). In practice, the signal covariance matrix also is unknown and is replaced by an estimate, the sample covariance matrix, i.e.,

$$\widehat{\mathbf{R}}_k = \frac{1}{G} \sum_{n=0}^{G-1} \mathbf{x}_k(n) \mathbf{x}_k^H(n). \tag{2.13}$$

There are some inherent tradeoffs in the choices of N and M and thereby G. The number of observations N should be chosen appropriately such that the signal $x_k(n)$ can be assumed to be stationary, while M should be chosen such that all significant non-zero correlations in the covariance matrix are modeled. On the other hand, M should also not be chosen higher than strictly necessary, since the goodness of the signal covariance matrix estimate in (2.13) depends on G being as high as possible. Additionally, since the evaluation of the likelihood function requires the existence (and calculation) of the inverse of the noise covariance matrix, it is required that rank $(\mathbf{Q}_k) = M$. This in turn implies that $G \geq M$ and hence $M < \frac{N}{2} + 1$. For a particular candidate fundamental frequency, we can construct the Vandermonde matrix $\mathbf{Z}_k$ in (2.11). However, we also need an estimate of the sinusoidal amplitudes in $\mathbf{P}_k$ to obtain a noise covariance matrix estimate. An obvious approach is to use the estimated signal covariance matrix instead of the noise covariance matrix resulting in a Capon-like amplitude estimator or one of the other amplitude estimates discussed in Chapter 5. A simple and asymptotically efficient estimate (see [155]) of the complex amplitudes for approximating the likelihood for various fundamental frequencies and orders is

$$\hat{\mathbf{a}}_k = \left(\sum_{n=0}^{G-1} \mathbf{Z}_k^H(n)\mathbf{Z}_k(n) \right)^{-1} \sum_{n=0}^{G-1} \mathbf{Z}_k^H(n)\mathbf{x}_k(n), \tag{2.14}$$

which can be approximated for large N as

$$\hat{\mathbf{a}}_k \approx \frac{1}{MG} \sum_{n=0}^{G-1} \mathbf{Z}_k^H(n)\mathbf{x}_k(n). \tag{2.15}$$

This is a number of phase-shifted fast Fourier transforms (FFTs), since $\mathbf{Z}_k(n)$ can be written as

$$\mathbf{Z}_k(n) = \mathbf{Z}_k \begin{bmatrix} e^{j\omega_k n} & & 0 \\ & \ddots & \\ 0 & & e^{j\omega_k L_k n} \end{bmatrix}. \tag{2.16}$$

Having found the complex amplitudes associated with a certain candidate fundamental frequency, we can now estimate the noise covariance matrix from the order L_k sinusoidal model as

$$\widehat{\mathbf{Q}}_k = \widehat{\mathbf{R}}_k - \mathbf{Z}_k\widehat{\mathbf{P}}_k\mathbf{Z}_k^H = \widehat{\mathbf{R}}_k - \sum_{l=1}^{L_k} \hat{A}_{k,l}^2 \mathbf{z}_l\mathbf{z}_l^H, \tag{2.17}$$

with $\widehat{\mathbf{P}}_k = \text{diag}\left(\begin{bmatrix} \hat{A}_{k,1}^2 & \cdots & \hat{A}_{k,l}^2 \end{bmatrix} \right)$. The inverse noise covariance matrix may be needed for and direct inversion for different model orders and fundamental frequencies poses a significant computational burden. Therefore, it is advantageous to compute it using the matrix inversion lemma. This leads to the following iterative expression for the inverse covariance matrix for $l = 1, 2, \ldots, L_k$:

$$\left(\widehat{\mathbf{Q}}_k^{(l)} \right)^{-1} = \left(\widehat{\mathbf{Q}}_k^{(l-1)} \right)^{-1} + \left(\widehat{\mathbf{Q}}_k^{(l-1)} \right)^{-1} \frac{\hat{A}_{k,l}^2 \mathbf{z}_l\mathbf{z}_l^H}{1 - \hat{A}_{k,l}^2 \mathbf{z}_l^H \left(\widehat{\mathbf{Q}}_k^{(l-1)} \right)^{-1} \mathbf{z}_l} \left(\widehat{\mathbf{Q}}_k^{(l-1)} \right)^{-1}, \tag{2.18}$$

with $\left(\widehat{\mathbf{Q}}_k^{(0)}\right)^{-1} = \widehat{\mathbf{R}}_k^{-1}$. Similarly, the matrix determinant lemma can be used for computing the determinant of $\mathbf{Q}_k^{(l)}$ in iteration l recursively as

$$\det\left(\widehat{\mathbf{Q}}_k^{(l)}\right) = \det\left(\widehat{\mathbf{Q}}_k^{(l-1)} - \hat{A}_{k,l}^2 \mathbf{z}_l \mathbf{z}_l^H\right) \tag{2.19}$$

$$= \left(1 - \hat{A}_{k,l}^2 \mathbf{z}_l^H \left(\widehat{\mathbf{Q}}_k^{(l-1)}\right)^{-1} \mathbf{z}_l\right) \det\left(\widehat{\mathbf{Q}}_k^{(l-1)}\right). \tag{2.20}$$

It must be stressed that the amplitude estimates used in updating the determinant and the inverse noise covariance matrix estimates should be chosen with care, since particular estimates may lead to rank deficiency in some cases. It can be seen that the likelihood function contains the term $\sum_{n=0}^{G-1} \mathbf{e}_k^H(n)\mathbf{Q}_k^{-1}\mathbf{e}_k(n)$. If we replace the vectors $\mathbf{e}_k(n)$ by estimates obtained as

$$\hat{\mathbf{e}}_k(n) = \mathbf{x}_k(n) - \hat{\mathbf{s}}_k(n), \tag{2.21}$$

and the noise covariance matrix by its estimate $\widehat{\mathbf{Q}}_k$, we can write the term as

$$\sum_{n=0}^{G-1} \hat{\mathbf{e}}_k^H(n)\widehat{\mathbf{Q}}_k^{-1}\hat{\mathbf{e}}_k(n) = \mathrm{Tr}\left\{\sum_{n=0}^{G-1} \hat{\mathbf{e}}_k^H(n)\widehat{\mathbf{Q}}_k^{-1}\hat{\mathbf{e}}_k(n)\right\} \tag{2.22}$$

$$= \mathrm{Tr}\left\{\widehat{\mathbf{Q}}_k^{-1} \sum_{n=0}^{G-1} \hat{\mathbf{e}}_k(n)\hat{\mathbf{e}}_k^H(n)\right\}. \tag{2.23}$$

Realizing that $\sum_{n=0}^{G-1} \hat{\mathbf{e}}_k(n)\hat{\mathbf{e}}_k^H(n)$ is the scaled sample noise covariance matrix obtained using the estimated deterministic signal components $\hat{\mathbf{s}}_k(n)$, we get that

$$\sum_{n=0}^{G-1} \hat{\mathbf{e}}_k^H(n)\widehat{\mathbf{Q}}_k^{-1}\hat{\mathbf{e}}_k(n) = \mathrm{Tr}\left\{\widehat{\mathbf{Q}}_k^{-1}\widehat{\mathbf{Q}}_k G\right\} = MG, \tag{2.24}$$

which means that this term is constant. The likelihood function therefore only depends on the determinant of $\widehat{\mathbf{Q}}_k$, but we can make use of the inverse noise covariance update formula for efficient computation of this quantity so our efforts have not been futile. It should be noted that the matrix obtained from $\sum_{n=0}^{G-1} \hat{\mathbf{e}}_k(n)\hat{\mathbf{e}}_k^H(n)$ may not be the exact inverse of the matrix obtained using (2.18), since the latter is based on $\mathbf{P}_k$ being diagonal.

An alternative approach to pitch estimation in the presence of colored noise is the method of [48], which is also based on maximum likelihood principles.

2.4   WHITE NOISE CASE

In some cases, the stochastic part of the signal model in (1.1) is white or approximately so. Not only can the computational complexity of the maximum likelihood estimator be reduced significantly, but the inherent sub-optimality of covariance matrix-based methods can be avoided. For the white noise case, the structure of the noise covariance matrix is now known, i.e., it reduces to a scaled diagonal matrix $\mathbf{Q}_k = \sigma_k^2 \mathbf{I}$ where σ_k is the variance of the noise $e_k(n)$. This has the consequence that we no longer need to estimate a full covariance matrix but only the variance, and, therefore, there is no need to split the observed signal into sub-vectors, i.e., we can simply set $M = N$ and thus $G = 1$. For notational simplicity, we define $\hat{\mathbf{e}}_k(0) = \hat{\mathbf{e}}_k$ and $\mathbf{x}_k(0) = \mathbf{x}_k$. The log-likelihood function can now be written as

$$\mathcal{L}(\boldsymbol{\theta}_k) = -N \ln \pi - N \ln \sigma_k^2 - \frac{1}{\sigma_k^2} \|\hat{\mathbf{e}}_k\|_2^2. \tag{2.25}$$

It can be seen that the maximum likelihood estimator (MLE) is simply the minimizer of the 2-norm of the modeling error $\hat{\mathbf{e}}_k$. Since the fundamental frequency is a nonlinear parameter, the method based on minimizing the 2-norm is called the nonlinear least-squares (NLS) method. The noise variance is generally unknown too, and we need to form an estimate of it to evaluate the log-likelihood, and, like the noise covariance matrix estimate, this estimate will depend on the order L_k. For a particular fundamental frequency candidate, assuming L_k harmonics, the maximum likelihood noise variance estimate is

$$\hat{\sigma}_k^2 = \frac{1}{N} \|\mathbf{x}_k - \boldsymbol{\Pi}_Z \mathbf{x}_k\|_2^2, \tag{2.26}$$

where $\boldsymbol{\Pi}_{Z_k}$ is the projection matrix defined as

$$\boldsymbol{\Pi}_{Z_k} = \mathbf{Z}_k \left(\mathbf{Z}_k^H \mathbf{Z}_k \right)^{-1} \mathbf{Z}_k^H. \tag{2.27}$$

The estimate in (2.26) is basically obtained by replacing the amplitudes by least-squares estimates. We note that, given the frequencies of the harmonics, the least-squares amplitude estimates can be obtained in an efficient manner by a recursive implementation [39]. Complex sinusoids are asymptotically orthogonal for any set of distinct frequencies, which means that the projection matrix $\boldsymbol{\Pi}_{Z_k}$ can be approximated as

$$\lim_{N \to \infty} N \boldsymbol{\Pi}_{Z_k} = \lim_{N \to \infty} N \mathbf{Z}_k \left(\mathbf{Z}_k^H \mathbf{Z}_k \right)^{-1} \mathbf{Z}_k^H \tag{2.28}$$
$$= \mathbf{Z}_k \mathbf{Z}_k^H. \tag{2.29}$$

Here, it should be stressed, that for a given N, this approximation will get progressively worse as ω_k gets smaller and it may be quite poor for low ω_k. Using the approximation in (2.29), the noise variance estimate can be simplified, i.e.,

$$\hat{\sigma}_k^2 \approx \frac{1}{N} \|\mathbf{x}_k - \frac{1}{N} \mathbf{Z}_k \mathbf{Z}_k^H \mathbf{x}_k\|_2^2. \tag{2.30}$$

We then get the following log-likelihood function, which depends only on ω_k by inserting this variance estimate into (2.25),

$$\mathcal{L}(\omega_k) \approx -N \ln \pi - N \ln \hat{\sigma}_k^2 - N. \tag{2.31}$$

We see that the amplitude can be found for various fundamental frequencies and orders using one FFT, this time, however, we are ignoring the noise color. From (2.31), we can observe a problem in relation to the model order. As the order L_k is increased, the log-likelihood is increased and therefore the maximum likelihood estimator will lead to the choice of the highest possible order. This phenomenon is illustrated in Figure 2.1 where (2.31) is shown as a function of the model order for a synthetic signal with $L_k = 5$. We will return to this problem and its solution later on.

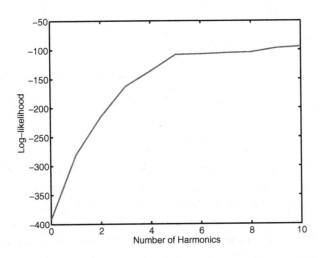

Figure 2.1: Typical log-likelihood function as a function of the model order for a synthetic periodic signal with the true order being five in white Gaussian noise.

Writing out the log-likelihood function, we see that for a given model order L_k the estimator reduces to

$$\hat{\omega}_k = \arg\max_{\omega_k} \mathbf{x}_k^H \mathbf{\Pi}_{Z_k} \mathbf{x}_k \tag{2.32}$$

$$\approx \arg\max_{\omega_k} \mathbf{x}_k^H \mathbf{Z}_k \mathbf{Z}_k^H \mathbf{x}_k \tag{2.33}$$

$$= \arg\max_{\omega_k} \|\mathbf{Z}_k^H \mathbf{x}_k\|_2^2, \tag{2.34}$$

where the last two lines follow from the asymptotic approximation, while the first line is exact. In practice, the estimator can be implemented by first evaluating the cost function on a coarse grid from which an initial estimate can be obtained. This initial estimate can then be refined using various numerical optimization techniques such as steepest descent and Newton's method (see, e.g., [6]). For the specific estimator considered here, such methods can be found in [33].

The estimator in (2.34) can be implemented efficiently by forming such a coarse initial estimate using an FFT. To see this, we introduce the Fourier transform of the observed signal $x_k(n)$ as

$$X_k(\omega) = \sum_{n=0}^{N-1} x_k(n)e^{-j\omega n}, \tag{2.35}$$

from which we see that the cost function can be expressed as

$$\|\mathbf{Z}_k^H \mathbf{x}_k\|_2^2 = \sum_{l=1}^{L_k} |\sum_{n=0}^{N-1} x_k(n)e^{-j\omega_k ln}|^2 \tag{2.36}$$

$$= \sum_{l=1}^{L_k} |X_k(\omega_k l)|^2. \tag{2.37}$$

The quantity $|X(\omega_k l)|^2$ is known in estimation theoretical terms as the periodogram (except for a scale factor) when obtained from finite length data records. It can be seen that the estimator reduces to a summation over the periodogram estimate of the power spectral density evaluated at the harmonic frequencies. This shows that the so-called sub-harmonic summation method [74] and the harmonic sum spectrum method [124] are asymptotically identical to the maximum likelihood estimator, whose exact version has previously been proposed in [168]. The method of [136] is essentially the same estimator, although modified to account for colored noise (on a historical note, the interested reader may also wish to consult [70]). In Figure 2.2, the cost function (2.37) is plotted for two different models orders for the noisy signal in Figure 1.2 having the spectrum shown in Figure 1.3. A number of things can be observed from the figure. First, for the true model order, $L_k = 5$, the cost function has a global maximum for the true fundamental at $\omega_k = 0.3142$, but for $L_k = 8$ the global maximum is at half the true fundamental. This clearly shows the importance of using the correct order. Second, it can also be seen that the cost functions are multi-modal, meaning that they have multiple extrema. These extrema are all quite narrow making numerical optimization difficult for large N. In Figure 2.3, the cost functions for the two signals in Figures 1.1 and 1.4 are shown.

About the use of the asymptotic approximations, it should be noted that these may not be very accurate for speech signals while this is often less of an issue for music signals, as music signals frequently can be assumed stationary over longer time intervals. As we have already stated, the approximation becomes increasingly worse the less separated the harmonics are in frequency. A change of sampling frequency does not generally alleviate this problem, as the normalized frequency is lowered by an increase in the sampling frequency. However, the CRLB does suggest that, with respect to the influence of the noise, it is, in fact, beneficial to use a high sampling frequency.

It is straight-forward to incorporate the parametric inharmonicity model of equation (1.28) in these estimators. Consider, for example, the cost function in (2.37). To apply the model of (1.28),

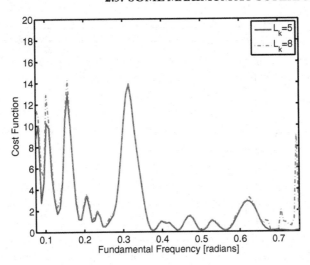

Figure 2.2: Approximate maximum likelihood cost function (2.37) evaluated for different model order for a synthetic signal having a fundamental frequency of 0.3142. The estimates can be identified as the largest peak.

we only have to replace the frequencies $\omega_k l$ by $\psi_{k,l} = \omega_k l \sqrt{1 + B_k l^2}$, i.e., the estimator becomes

$$(\hat{\omega}_k, \hat{B}_k) = \arg \max_{\omega_k, B_k} \sum_{l=1}^{L_k} |X_k(\psi_{k,l})|^2 \tag{2.38}$$

$$= \arg \max_{\omega_k, B_k} \sum_{l=1}^{L_k} |X_k(\omega_k l \sqrt{1 + B_k l^2})|^2, \tag{2.39}$$

which means that we, in principle, have to sweep over combinations of the two nonlinear parameters to obtain the estimates.

It is worth noting that, because the white noise case is so much simpler than the colored case, the estimator presented here may be preferred over the colored noise estimator in real-time applications even if the noise is known not to be completely white, as it may still provide adequately accurate estimates, especially so if the model order is known, since the nonlinear least-squares method is asymptotically efficient for the parameters of the individual harmonics even for colored noise [153].

2.5   SOME MAXIMUM A POSTERIORI ESTIMATORS

An alternative to maximum likelihood estimation is maximum a posteriori (MAP) estimation, where the so-called posterior PDF is maximized, i.e.,

$$\hat{\boldsymbol{\theta}}_k = \arg \max_{\boldsymbol{\theta}_k} p(\boldsymbol{\theta}_k | \mathbf{x}_k). \tag{2.40}$$

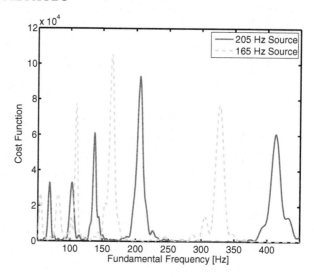

Figure 2.3: Approximate maximum likelihood cost function for the two signals in Figures 1.1 and 1.4, respectively.

This means that the parameters that are most likely, given the observation $\mathbf{x}_k$, are chosen as the estimates. Bayes' rule states that we may rewrite the posterior PDF as

$$p(\boldsymbol{\theta}_k|\mathbf{x}_k) = \frac{p(\mathbf{x}_k|\boldsymbol{\theta}_k)p(\boldsymbol{\theta}_k)}{p(\mathbf{x}_k)}, \tag{2.41}$$

and, since $p(\mathbf{x}_k)$ is constant once $\mathbf{x}_k$ has been observed, we may maximize $p(\mathbf{x}_k|\boldsymbol{\theta}_k)p(\boldsymbol{\theta}_k)$ instead of (2.40), or, more conveniently, its logarithm, i.e.,

$$\hat{\boldsymbol{\theta}}_k = \arg\max_{\boldsymbol{\theta}_k} \ln p(\mathbf{x}_k|\boldsymbol{\theta}_k) + \ln p(\boldsymbol{\theta}_k). \tag{2.42}$$

The density $p(\boldsymbol{\theta}_k)$ is referred to as a prior on $\boldsymbol{\theta}_k$ as it incorporates prior information on the distribution of the parameters, and this is exactly the strength of this approach. The estimator can be seen to be comprised of two terms: a log-likelihood term and a log-prior term; in the case where $p(\boldsymbol{\theta}_k)$ is constant, the method essentially reduces to the maximum likelihood estimator.

As an example of its applications to pitch estimation consider the following: it has been noted by many authors that the frequencies of the fundamental frequencies in speech and audio signals are not uniformly distributed over the whole interval and this can be taken into account using the MAP approach. Similarly, one may argue that the tones of music are more likely to occur at fundamental frequencies equivalent to the tones of musical scales or tunings. There are, in other words, a myriad of possible applications of this principle, but we will here consider only two.

First, we will consider the case of incorporating prior knowledge of the distribution of the fundamental frequency in our estimation framework. Recall that the log-likelihood function of the observed signal for the case of white Gaussian noise is given by

$$\ln p(\mathbf{x}_k|\omega_k) = -N \ln \pi - N \ln \hat{\sigma}_k^2 - N. \tag{2.43}$$

To be able to make use of the general MAP estimator in (2.42), we assume uniform priors on the amplitudes and the noise variance, and that these parameters are mutually independent and independent of the fundamental frequency. In that case, we can substitute the variance estimate in (2.43) by

$$\hat{\sigma}_k^2 = \frac{1}{N}\|\mathbf{x}_k - \mathbf{Z}_k\left(\mathbf{Z}_k^H \mathbf{Z}_k\right)^{-1}\mathbf{Z}_k^H \mathbf{x}_k\|_2^2. \tag{2.44}$$

Next, we introduce a log-prior $\ln p(\omega_k)$ on the fundamental frequency. By combining these expression with the estimator in (2.42) and dropping all constant terms, we obtain

$$\hat{\omega}_k = \arg\max_{\omega_k} -N \ln \hat{\sigma}_k^2 + \ln p(\omega_k) \tag{2.45}$$

$$= \arg\max_{\omega_k} -N \ln \|\mathbf{x}_k - \mathbf{Z}_k\left(\mathbf{Z}_k^H \mathbf{Z}_k\right)^{-1}\mathbf{Z}_k^H \mathbf{x}_k\|_2^2 + \ln p(\omega_k). \tag{2.46}$$

By writing out the 2-norm as in (2.34), we get

$$\hat{\omega}_k = \arg\max_{\omega_k} -N \ln \|\mathbf{x}_k - \mathbf{Z}_k\left(\mathbf{Z}_k^H \mathbf{Z}_k\right)^{-1}\mathbf{Z}_k^H \mathbf{x}_k\|_2^2 + \ln p(\omega_k) \tag{2.47}$$

$$\approx \arg\max_{\omega_k} -N \ln \left(\mathbf{x}_k^H \mathbf{x}_k - \frac{1}{N}\mathbf{x}_k^H \mathbf{Z}_k \mathbf{Z}_k^H \mathbf{x}_k\right) + \ln p(\omega_k), \tag{2.48}$$

where we have used the asymptotic properties of the Vandermonde matrix like in (2.29) to obtain the last line. Using Parseval's relation, the estimator can also be expressed in the frequency domain as

$$\hat{\omega}_k = \arg\max_{\omega_k} -N \ln \left(\frac{1}{2\pi}\int_{-\pi}^{\pi}|X_k(\psi)|^2 d\psi - \frac{1}{N}\sum_{l=1}^{L_k}|X_k(\omega_k l)|^2\right) + \ln p(\omega_k). \tag{2.49}$$

It can be seen from (2.49) that the log-prior acts like a penalty term in the optimization problem associated with the estimation problem. For most speech and audio signals, a good log-prior would be one that emphasizes low frequencies. Priors that would achieve this could be Laplacian, Rayleigh or Maxwell distributions.

We proceed to examine the perturbed signal model introduced in (1.29) where the frequencies of the harmonics are allowed to deviate from being integer multiples of the fundamental. More specifically, we will use the following signal model

$$x_k(n) = \sum_{l=1}^{L_k} a_{k,l} e^{j(\omega_k l + \Delta_{k,l})n} + e_k(n) \tag{2.50}$$

in combination with the MAP approach to solve the problem of finding the fundamental frequency. We will assume uniform priors and independence of the fundamental frequency and the amplitudes. Furthermore, we assume that the noise variance is known. The log-likelihood function is then given by

$$\ln p(\mathbf{x}_k|\omega_k, \{\Delta_{k,l}\}) = -N \ln \pi - N \ln \sigma_k^2 - \frac{1}{\sigma_k^2}\|\hat{\mathbf{e}}_k\|_2^2, \tag{2.51}$$

where

$$\hat{\mathbf{e}}_k = \mathbf{x}_k - \mathbf{Z}_k \left(\mathbf{Z}_k^H \mathbf{Z}_k\right)^{-1} \mathbf{Z}_k^H \mathbf{x}_k. \tag{2.52}$$

To be able to solve the problem of finding the fundamental frequency of (2.50), we need to impose some restrictions on the perturbations, $\{\Delta_{k,l}\}$. Specifically, if we allow these to become arbitrarily large, any ω_k will fit. Therefore, the traditional ML method cannot be used. We expect that most of the perturbations will be rather small, but also that a few large ones may occur. A PDF that describes such distributions is the Laplacian PDF and this is what we will use here. Assuming that the perturbations are independent, their log-prior is given by

$$\ln p(\{\Delta_{k,l}\}) = -L_k \ln 2\gamma - \frac{1}{\gamma}\sum_{l=1}^{L_k} |\Delta_{k,l}| \tag{2.53}$$

$$= -L_k \ln 2\gamma - \frac{1}{\gamma}\|\mathbf{\Delta}_k\|_1, \tag{2.54}$$

where $\mathbf{\Delta}_k = [\ \Delta_{k,1} \ \dots \ \Delta_{k,L_k}\ ]$ and $2\gamma^2$ is the variance with $\gamma > 0$. By combining (2.51) and (2.54) using (2.42), we obtain the following estimator:

$$\left(\hat{\omega}_k, \{\hat{\Delta}_{k,l}\}\right) = \arg\max_{\omega_k,\{\Delta_{k,l}\}} \ln p(\mathbf{x}_k|\omega_k, \{\Delta_{k,l}\}) + \ln p(\{\Delta_{k,l}\}) \tag{2.55}$$

$$= \arg\max_{\omega_k,\{\Delta_{k,l}\}} -N\ln\pi - N\ln\sigma_k^2 - \frac{1}{\sigma_k^2}\|\hat{\mathbf{e}}_k\|_2^2 - L_k \ln 2\gamma - \frac{1}{\gamma}\|\mathbf{\Delta}_k\|_1$$

$$= \arg\min_{\omega_k,\{\Delta_{k,l}\}} \frac{1}{\sigma_k^2}\|\hat{\mathbf{e}}_k\|_2^2 + \frac{1}{\gamma}\|\mathbf{\Delta}_k\|_1 \tag{2.56}$$

$$= \arg\min_{\omega_k,\{\Delta_{k,l}\}} \|\hat{\mathbf{e}}_k\|_2^2 + \frac{\sigma_k^2}{\gamma}\|\mathbf{\Delta}_k\|_1. \tag{2.57}$$

This is a joint nonlinear optimization problem over the fundamental frequency and all the perturbations. Regrettably, it appears to be nontrivial to solve this problem. An optimization problem of the form in (2.57) is also known as a regularized optimization where the last term is referred to as the regularization term and σ_k^2/γ as the regularization constant. It is worth noting that the regularization constant depends not only on the variance of the perturbations but also on the noise variance.

Using the same asymptotic approximations as in (2.49), we can simplify the estimator somewhat, i.e.,

$$\left(\hat{\omega}_k, \{\hat{\Delta}_{k,l}\}\right) = \arg \min_{\omega_k, \{\Delta_{k,l}\}} \|\hat{\mathbf{e}}_k\|_2^2 + \frac{\sigma_k^2}{\gamma} \|\mathbf{\Delta}_k\|_1 \tag{2.58}$$

$$= \arg \max_{\omega_k, \{\Delta_{k,l}\}} \frac{1}{N} \sum_{l=1}^{L_k} |X_k(\omega_k l + \Delta_l)|^2 - \frac{\sigma_k^2}{\gamma} \sum_{l=1}^{L_k} |\Delta_{k,l}|. \tag{2.59}$$

It is perhaps still not clear how to solve this problem, but by substituting $\Delta_{k,l}$ by $\psi_{k,l} - \omega_k l$ and $\omega_k l + \Delta_{k,l}$ by $\psi_{k,l}$ and observing that the inner multidimensional nonlinear optimization problem can be decoupled into a number of one-dimensional problems, we obtain

$$\hat{\omega}_k = \arg \max_{\omega_k} \max_{\{\Delta_{k,l}\}} \frac{1}{N} \sum_{l=1}^{L_k} |X_k(\omega_k l + \Delta_{k,l})|^2 - \frac{\sigma_k^2}{\gamma} \sum_{l=1}^{L_k} |\Delta_{k,l}| \tag{2.60}$$

$$= \arg \max_{\omega_k} \max_{\{\psi_{k,l}\}} \frac{1}{N} \sum_{l=1}^{L_k} |X_k(\psi_{k,l})|^2 - \frac{\sigma_k^2}{\gamma} \sum_{l=1}^{L_k} |\psi_{k,l} - \omega_k l| \tag{2.61}$$

$$= \arg \max_{\omega_k} \frac{1}{N} \sum_{l=1}^{L_k} \max_{\psi_{k,l}} \left(|X_k(\psi_{k,l})|^2 - \frac{\sigma_k^2}{\gamma} |\psi_{k,l} - \omega_k l| \right). \tag{2.62}$$

The last line shows that it is actually possible to solve the problem for a given candidate fundamental frequency by simply finding the peak of a modified periodogram $|X_k(\psi_{k,l})|^2$ where a penalty term has been added. The good thing is that the periodogram only has to be calculated once and the rest is a matter of adding different penalty terms. Once the fundamental frequency has been determined, the frequencies of the individual harmonics can be obtained as

$$\hat{\psi}_{k,l} = \arg \max_{\psi_{k,l}} \left(|X_k(\psi_{k,l})|^2 - \frac{\sigma_k^2}{\gamma} |\psi_{k,l} - \hat{\omega}_k l| \right) \tag{2.63}$$

and the perturbations as $\hat{\Delta}_{k,l} = \hat{\omega}_k l - \hat{\psi}_{k,l}$.

It is curious to note that the estimators in (2.49) and (2.62) are different. Suppose that we had used a similar prior on ω_k in (2.49) as in (2.54). Then, (2.49) would consist of the logarithm of the error norm and a regularization term consisting of merely a norm, as opposed to (2.62) where both terms are simply scaled vector norms. Interestingly, the estimator in (2.62) is of the form usually referred to as a regularized cost function in optimization, which is sometimes justified as a MAP-like estimator. Also, the basic idea of the estimator proposed in [92], i.e., that amplitudes are usually smooth over frequency, can also be cast in the framework of MAP estimation by introducing a prior on the differences between the real amplitudes of adjacent harmonics. This can be realized by introducing a so-called *smoothing regularization* term [14], although some additional work is required, since this prior applies to the real amplitudes and not the complex ones.

2.6   MAP MODEL AND ORDER SELECTION

We will now address the problem of choosing between different models and different model orders from a statistical perspective. The problem of selecting between different model orders is a special case of the general model selection problem, where the models under consideration are nested, meaning that the simple models are special cases of the more complicated models. As has been discussed, the problem of determining the model order and the number of sources is the key to solving the pitch and multi-pitch estimation problems using parametric models. The model selection problem can, for example, be used for determining which of the usual harmonic model or the inharmonicity model should be used for a particular signal, whether there is just noise present and for taking missing harmonics into account. The most commonly used methods for order selection are the Akaike information criterion (AIC) [2] and the minimum description length criterion (MDL) [142, 143] (see also [151]). These method comprise a "goodness of fit" term and an order dependent penalty term that penalizes more complex models. It is easy to see why the penalty term is needed: a more complex model is generally capable of fitting the data better than a simple one, even though the simple model is the one that was used for generating the data. It was shown in [44] that linear and nonlinear parameters should not be given equal weight, something which is easy to see for the frequency estimation problem.

   The subject of model and order selection is worthy of an entire book on its own, and we will here only present some results and some brief arguments. More specifically, will derive model and order selection criteria using the MAP approach of [43, 44] (see also [160]). For simplicity, we will do this for the single-pitch case, i.e., based on $\mathbf{x}_k$ even though the principles can be used for multi-pitch signals too, for example in combination with some of the methods considered later in this chapter. First, we introduce $\mathbb{Z}_q = \{0, 1, \ldots, q - 1\}$ which is the candidate model index set with $\mathcal{M}_m, m \in \mathbb{Z}_q$ being the candidate models. Examples of candidate models that we would like to be able to choose from are: white Gaussian noise, colored Gaussian noise, various numbers of harmonically related sinusoids, sinusoids where the harmonics are not exact integer multiples of the fundamental and so forth. The principle of MAP-based model selection is to choose the model that maximizes the a posteriori probability of the model given the observation $\mathbf{x}_k$. It is easy to see that this can be written as

$$\widehat{\mathcal{M}}_k = \arg \max_{\mathcal{M}_m, m \in \mathbb{Z}_q} p(\mathcal{M}_m | \mathbf{x}_k) \qquad (2.64)$$

$$= \arg \max_{\mathcal{M}_m, m \in \mathbb{Z}_q} \frac{p(\mathbf{x}_k | \mathcal{M}_m) p(\mathcal{M}_m)}{p(\mathbf{x}_k)}. \qquad (2.65)$$

Assuming that all the models are equally probable, i.e.,

$$p(\mathcal{M}_m) = \frac{1}{q} \qquad (2.66)$$

and noting that $p(\mathbf{x}_k)$ is constant once $\mathbf{x}_k$ has been observed, the MAP model selection criterion reduces to

$$\widehat{\mathcal{M}}_k = \arg \max_{\mathcal{M}_m, m \in \mathbb{Z}_q} p(\mathbf{x}_k|\mathcal{M}_m), \tag{2.67}$$

which is the likelihood function when seen as a function of $\mathcal{M}_m$. Since the various models also depend on a number of unknown parameters, like amplitudes, phases, fundamental frequencies and so forth, here denoted $\boldsymbol{\theta}_k$, we will integrate those out as

$$p(\mathbf{x}_k|\mathcal{M}_m) = \int_{\boldsymbol{\Theta}_k} p(\mathbf{x}_k|\boldsymbol{\theta}_k, \mathcal{M}_m) p(\boldsymbol{\theta}_k|\mathcal{M}_m) d\boldsymbol{\theta}_k. \tag{2.68}$$

Note that the parameter vector $\boldsymbol{\theta}_k$ depends on the model $\mathcal{M}_m$, but we have here omitted this dependency in the notation. It is not generally possible to obtain any simple analytic expression for the integral in (2.68), especially for complicated nonlinear models such as ours, and we must seek another way of evaluating this integral. One such way is numerical integration, but we will here instead use the approach of [43, 44] based on the method of Laplace integration. Assuming that the likelihood function is highly peaked around maximum likelihood estimates $\hat{\boldsymbol{\theta}}_k$, which is usually the case for high N, (2.68) can be written as [43, 44] (see also [160])

$$\int_{\boldsymbol{\Theta}_k} p(\mathbf{x}_k|\boldsymbol{\theta}_k, \mathcal{M}_m) p(\boldsymbol{\theta}_k|\mathcal{M}_m) d\boldsymbol{\theta}_k = (2\pi)^{D_k/2} \det\left(\widehat{\mathbf{H}}_k\right)^{-1/2} p(\mathbf{x}_k|\hat{\boldsymbol{\theta}}_k, \mathcal{M}_m) p(\hat{\boldsymbol{\theta}}_k|\mathcal{M}_m), \tag{2.69}$$

where D_k is the number of parameters and

$$\widehat{\mathbf{H}}_k = - \left. \frac{\partial^2 \ln p(\mathbf{x}_k|\boldsymbol{\theta}_k, \mathcal{M}_m)}{\partial \boldsymbol{\theta}_k \partial \boldsymbol{\theta}_k^T} \right|_{\boldsymbol{\theta}_k = \hat{\boldsymbol{\theta}}_k} \tag{2.70}$$

is the Hessian of the log-likelihood function evaluated at $\hat{\boldsymbol{\theta}}_k$. Note that all parameters are real-valued and that the expression for the real multivariate Gaussian PDF is therefore used. Taking the logarithm of the right-hand side of (2.69) and sticking to tradition and ignoring terms of order $\mathcal{O}(1)$ and $\frac{D_k}{2} \ln(2\pi)$, which is negligible for large N, we get

$$\widehat{\mathcal{M}}_k = \arg \min_{\mathcal{M}_m, m \in \mathbb{Z}_q} \underbrace{- \ln p(\mathbf{x}_k|\hat{\boldsymbol{\theta}}_k, \mathcal{M}_m)}_{\text{log-likelihood}} + \underbrace{\frac{1}{2} \ln \det\left(\widehat{\mathbf{H}}_k\right)}_{\text{penalty}}, \tag{2.71}$$

which can be used directly for selecting between various models and orders. Moreover, a vague prior on $p(\boldsymbol{\theta}_k|\mathcal{M}_m)$ has been used to obtain (2.71) from (2.69) (see [43]). Now we will derive a criterion for selecting the model order of the single-pitch model in (1.1) and detecting the presence of a periodic source. To do this, observe that the Hessian defined in (2.70) is related to the Fischer information matrix (1.33) in a simple way. Only, it is evaluated in $\hat{\boldsymbol{\theta}}_k$ and no expectation is taken, and it is, therefore, often referred to as the observed information matrix. It was shown in [160] that

(2.70) can be used as an approximation (for large N) of the Fischer information matrix, and, hence, also vice versa, leading to the following approximation:

$$\widehat{\mathbf{H}}_k \approx -\mathrm{E} \left\{ \frac{\partial^2 \ln p(\mathbf{x}_k | \boldsymbol{\theta}_k)}{\partial \boldsymbol{\theta}_k \partial \boldsymbol{\theta}_k^T} \right\} \Bigg|_{\boldsymbol{\theta}_k = \hat{\boldsymbol{\theta}}_k} \tag{2.72}$$

The benefit of using (2.72) over (2.70) is that the former is readily available in the literature for most models, something that also is the case for our model. The diagonal terms are therefore given by (1.37)-(1.39) (the off-diagonal terms can be found in [23]), and we introduce the normalization matrix (see [160])

$$\mathbf{K}_N = \begin{bmatrix} N^{-3/2} & \mathbf{0} \\ \mathbf{O} & N^{-1/2}\mathbf{I} \end{bmatrix} \tag{2.73}$$

where $\mathbf{I}$ is an $2L_k \times 2L_k$ identity matrix. The first diagonal term is due to the fundamental frequency with the remainder being due to the L_k amplitudes and phases. Using this normalization matrix, we can write the determinant of the Hessian in (2.71) as

$$\det\left(\widehat{\mathbf{H}}_k\right) = \det\left(\mathbf{K}_N^{-2}\right) \det\left(\mathbf{K}_N \widehat{\mathbf{H}}_k \mathbf{K}_N\right). \tag{2.74}$$

And, finally, by observing that $\mathbf{K}_N \widehat{\mathbf{H}}_k \mathbf{K}_N = \mathcal{O}(1)$ and taking the logarithm of (2.74), we obtain

$$\ln \det\left(\widehat{\mathbf{H}}_k\right) = \ln \det\left(\mathbf{K}_N^{-2}\right) + \ln \det\left(\mathbf{K}_N \widehat{\mathbf{H}}_k \mathbf{K}_N\right) \tag{2.75}$$

$$= \ln \det\left(\mathbf{K}_N^{-2}\right) + \mathcal{O}(1) \tag{2.76}$$

$$= 3 \ln N + 2L_k \ln N + \mathcal{O}(1). \tag{2.77}$$

When the additive noise is a white complex Gaussian process, the log-likelihood function in (2.71) is $N \ln \sigma_k^2$ (see Section 2.4) where σ_k^2 then has to be replaced by an estimate for each candidate order L_k, denoted as $\hat{\sigma}_k^2(L_k)$. Finally, when substituting (2.77) into (2.71), we obtain the following simple and useful expression for selecting the model order:

$$\hat{L}_k = \arg \min_{L_k} \underbrace{N \ln \hat{\sigma}_k^2(L_k)}_{\text{log-likelihood}} + \underbrace{\frac{3}{2} \ln N + L_k \ln N}_{\text{penalty}} \tag{2.78}$$

To determine whether any harmonics are present at all, i.e., to perform pitch detection, the above cost function should be compared to the log-likelihood of the zero order model, meaning that no harmonics are present if

$$N \ln \hat{\sigma}_k^2(0) < N \ln \hat{\sigma}_k^2(\hat{L}_k) + \frac{3}{2} \ln N + \hat{L}_k \ln N, \tag{2.79}$$

where $\hat{\sigma}_k^2(0)$ is simply the variance of the observed signal. The rule in (2.79) is essentially a *pitch detection* rule as it detects the presence of a pitch. It can be seen that both (2.78) and (2.79) require the determination of the noise variance for each candidate model order. The criterion in (2.78) reflects the tradeoff between the variance of the residual and the complexity of the model. For example, for a high model order, the estimated variance will be low, but the number of parameters will be high. Conversely, for a low model order, there are only few parameters but a high variance residual. The MAP order estimation criterion in (2.78) is shown, for a given fundamental frequency, in Figure 2.4 along with the log-likelihood term $N \ln \hat{\sigma}_k^2(L_k)$ for a synthetic signal that consists of five harmonics. It can be seen that MAP criterion leads to identification of the correct model order. In Figure 2.5, fundamental frequency estimates obtained using a joint fundamental frequency and order estimator based on (2.78) and (2.37) are shown in the bottom panel. The estimates were obtained from 30 ms segments of the voiced speech signal having the spectrogram shown in the top panel. It can be seen, that even when the number of harmonics varies greatly, the estimator is still accurate. It can also be seen that the estimator is able to provide accurate estimates despite the rapid changes in pitch.

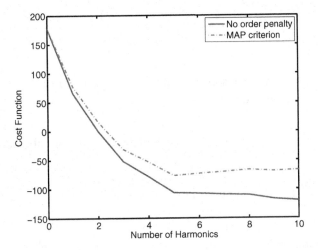

Figure 2.4: MAP model selection criterion and log-likelihood term for a synthetic signal with $L_k = 5$.

Similar criteria can be obtained for other signal models, like the perturbed model or the parametric inharmonicity model and the principles described here can be used for selecting between the various models in a signal-adaptive manner, but one has to derive the exact criterion for each such model. For linear models (in fact, for most models), the Hessian grows without bound as N increases ([44, 160]). In that case, the determinant of the Hessian can be simplified as

$$\det\left(\widehat{\mathbf{H}}_k\right) = \det\left(N\frac{1}{N}\widehat{\mathbf{H}}_k\right) \tag{2.80}$$

$$= \det(N\mathbf{I})\det\left(\frac{1}{N}\widehat{\mathbf{H}}_k\right), \tag{2.81}$$

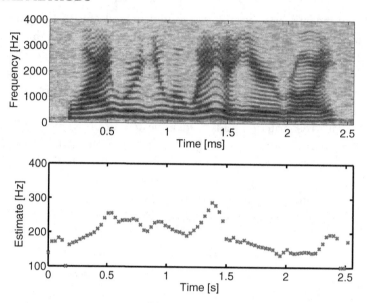

Figure 2.5: Voiced speech signal spectrogram (top) and pitch estimates obtained using a joint fundamental frequency and order estimator (bottom).

where the identity matrix is $D_k \times D_k$ (again with D_k being the number of real parameters). It then follows that

$$\ln \det \left(\widehat{\mathbf{H}}_k \right) = D_K \ln N + \mathcal{O}(1), \tag{2.82}$$

which can be used in combination with (2.71) to obtain order estimates.

In connection with the general problem of finding the number of sinusoids in noise using statistical methods, the interested reader may also want to consult [55, 56, 68, 87] on the matter. Later, we will show how subspace methods can be used for this.

2.7   FAST MULTI-PITCH ESTIMATION

We will now proceed to present a simple and computationally efficient method for multi-pitch estimation based on the principles considered in this chapter. As we have seen, the maximum likelihood estimator reduces to the minimizer of the 2-norm of the modeling error for white Gaussian noise. The fundamental frequency estimator based on this principle is what we refer to as the nonlinear least-squares (NLS) method. It turns out that the even when the noise is colored, the NLS method is still asymptotically efficient meaning that it attains the CRLB for large N [153]. This means that for finding the fundamental frequency, we need not worry about the color of the noise too much and this simplifies the problem significantly. However, it should be stressed, that for the problems of model and order selection the color of the noise is in fact important (see, e.g., [30] for a demonstration of this) as it affects both the likelihood term and the penalty term in the MAP method.

For convenience, we define a signal vector containing all N samples of the observed signal as $\mathbf{x} = \mathbf{x}(0)$ with $M = N$. The NLS estimates are obtained as the set of fundamental frequencies and amplitudes that minimize the 2-norm of the difference between this signal vector and the signal model, i.e.,

$$\{\hat{\omega}_k\} = \arg \min_{\{\mathbf{a}_k\},\{\omega_k\}} \left\| \mathbf{x} - \sum_{k=1}^{K} \mathbf{Z}_k \mathbf{a}_k \right\|_2^2 . \tag{2.83}$$

Assuming that all the frequencies in $\{\mathbf{Z}_k\}$ are distinct and well separated and that $N \gg 1$, (2.83) can be well-approximated by finding the fundamental frequency of the individual sources, i.e.,

$$\hat{\omega}_k = \arg \min_{\mathbf{a}_k,\omega_k} \|\mathbf{x} - \mathbf{Z}_k \mathbf{a}_k\|_2^2 . \tag{2.84}$$

The minimizations of (2.83) and (2.84) are equivalent only when the matrices $\{\mathbf{Z}_k\}$ are orthogonal. This is only the case for peculiar special cases for finite length signals, but is true asymptotically in N as long as no harmonics overlap. Therefore, the two cost functions can be expected to yield similar results for large N.

Minimizing (2.84) with respect to the complex amplitudes $\mathbf{a}_k$ gives the estimates $\hat{\mathbf{a}}_k = \left(\mathbf{Z}_k^H \mathbf{Z}_k\right)^{-1} \mathbf{Z}_k^H \mathbf{x}$, which, when inserted in (2.84), yields

$$\hat{\omega}_k = \arg \max_{\omega_k} \mathbf{x}^H \mathbf{Z}_k \left(\mathbf{Z}_k^H \mathbf{Z}_k\right)^{-1} \mathbf{Z}_k^H \mathbf{x} \tag{2.85}$$

$$\approx \arg \max_{\omega_k} \mathbf{x}^H \mathbf{Z}_k \mathbf{Z}_k^H \mathbf{x}, \tag{2.86}$$

where the last line follows from the assumption that $N \gg 1$. In Figure 2.6, the cost function of the estimator in (2.86) is shown for the mixture of the two signals in Figures 1.1 and 1.4 in additive white Gaussian noise. The fundamental frequencies can be identified as the two largest peaks, and it can be seen that the fundamental frequencies of the two sources, i.e., 205 Hz and 165 Hz, are, in fact, found using this method. However, it can also be seen that the cost function is extremely complicated containing multiple very sharp peaks.

The first estimator in (2.85) is exact for the single-pitch case where $\mathbf{x}$ is replaced by $\mathbf{x}_k$. The resulting cost function can be written as $\|\mathbf{Z}_k^H \mathbf{x}\|_2^2$ where the matrix product $\mathbf{Z}_k^H \mathbf{x}$ can be implemented efficiently for a linear grid search over ω_k using a FFT algorithm. An alternative interpretation of the approximate NLS estimator is as follows: the cost function can be written as $\sum_{l=1}^{L} \|\mathbf{z}(\omega_k l)^H \mathbf{x}\|_2^2$, which is the periodogram power spectral density estimate of $\mathbf{x}$ evaluated at and summed over the harmonic frequencies $\omega_k l$.

Interestingly, the cost function in (2.86) can be also be written as

$$\|\mathbf{Z}_k^H \mathbf{x}\|_2^2 = \text{Tr} \left[\mathbf{Z}_k^H \mathbf{x} \mathbf{x}^H \mathbf{Z}_k \right]. \tag{2.87}$$

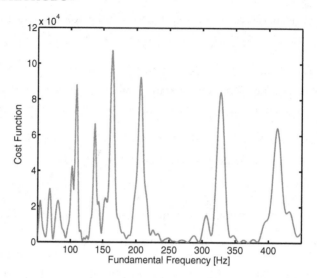

Figure 2.6: Approximate maximum likelihood cost function for the mixture of the two signals in Figures 1.1 and 1.4 having fundamental frequencies 205 and 165 Hz. Note that some noise has been added.

We can instead take the expected value after replacing $\mathbf{x}$ by the sub-vector $\mathbf{x}(n)$, with $M < N$, in (2.87), i.e.,

$$\mathrm{E}\left\{\|\mathbf{Z}_k^H \mathbf{x}(n)\|_2^2\right\} = \mathrm{Tr}\left[\mathbf{Z}_k^H \mathbf{R} \mathbf{Z}_k\right], \tag{2.88}$$

which results in the following estimator:

$$\hat{\omega}_k = \arg\max_{\omega_k} \mathrm{Tr}\left[\mathbf{Z}_k^H \mathbf{R} \mathbf{Z}_k\right]. \tag{2.89}$$

Instead of matching the signal model to a single snapshot of $\mathbf{x}$ as in (2.87), this estimator matches the model to the covariance matrix. Later, we will see that this leads to a filterbank interpretation of this method and an asymptotic relation to optimal filtering methods.

2.8   EXPECTATION MAXIMIZATION

The next algorithm is based on the Expectation Maximization (EM) algorithm [101], whose name is somewhat of a misnomer. The EM algorithm is an iterative method for maximum likelihood estimation involving several nonlinear parameters. The method presented here is based on [33], which may be considered a special case of [51]. More specifically, [51] deals with the general problem of estimating of the parameters of superimposed signals. The curious reader may also want to consult [159] for an alternative explanation of the EM algorithm. In our case, the superimposed signals are the periodic sources.

We will now prove the convergence of the EM algorithm following [111] before exploring the details for the problem at hand. The log-likelihood of the observed data is defined as

$$\mathcal{L}(\boldsymbol{\theta}) = \ln p(\mathbf{x}; \boldsymbol{\theta}). \tag{2.90}$$

We can rewrite this function by introducing the variable $\mathbf{y}$, which we will then integrate out, as

$$\ln p(\mathbf{x}; \boldsymbol{\theta}) = \ln \int p(\mathbf{x}, \mathbf{y}; \boldsymbol{\theta}) dy \tag{2.91}$$

$$= \ln \int p(\mathbf{x}, \mathbf{y}; \boldsymbol{\theta}) \frac{p(\mathbf{y}|\mathbf{x}; \hat{\boldsymbol{\theta}})}{p(\mathbf{y}|\mathbf{x}; \hat{\boldsymbol{\theta}})} dy, \tag{2.92}$$

where we have multiplied and divided by $p(\mathbf{y}|\mathbf{x}; \hat{\boldsymbol{\theta}})$ to obtain the last line. Next, we note that this can be interpreted as a conditional expectation over $\mathbf{y}$, i.e.,

$$\ln \int \frac{p(\mathbf{x}, \mathbf{y}; \boldsymbol{\theta})}{p(\mathbf{y}|\mathbf{x}; \hat{\boldsymbol{\theta}})} p(\mathbf{y}|\mathbf{x}; \hat{\boldsymbol{\theta}}) dy = \ln E_{\mathbf{y}} \left\{ \frac{p(\mathbf{x}, \mathbf{y}; \boldsymbol{\theta})}{p(\mathbf{y}|\mathbf{x}; \hat{\boldsymbol{\theta}})} \middle| \mathbf{x}; \hat{\boldsymbol{\theta}} \right\}, \tag{2.93}$$

where $E_{\mathbf{y}} \left\{ \cdot \middle| \mathbf{x}; \hat{\boldsymbol{\theta}} \right\}$ is the conditional expectation over $\mathbf{y}$ conditioned on $\mathbf{x}$. The key observation regarding the convergence of the EM algorithm follows from Jensen's inequality,, which states that $f(E_{\mathbf{x}}\{\mathbf{x}\}) \geq E_{\mathbf{x}}\{f(\mathbf{x})\}$ for a concave function $f(\mathbf{x})$. Therefore, we may rewrite the conditional expectation as

$$\ln E_{\mathbf{y}} \left\{ \frac{p(\mathbf{x}, \mathbf{y}; \boldsymbol{\theta})}{p(\mathbf{y}|\mathbf{x}; \hat{\boldsymbol{\theta}})} \middle| \mathbf{x}; \hat{\boldsymbol{\theta}} \right\} \geq E_{\mathbf{y}} \left\{ \ln \left(\frac{p(\mathbf{x}, \mathbf{y}; \boldsymbol{\theta})}{p(\mathbf{y}|\mathbf{x}; \hat{\boldsymbol{\theta}})} \right) \middle| \mathbf{x}; \hat{\boldsymbol{\theta}} \right\} \tag{2.94}$$

$$= \int \ln \left(\frac{p(\mathbf{x}, \mathbf{y}; \boldsymbol{\theta})}{p(\mathbf{y}|\mathbf{x}; \hat{\boldsymbol{\theta}})} \right) p(\mathbf{y}|\mathbf{x}; \hat{\boldsymbol{\theta}}) dy. \tag{2.95}$$

The right-hand side of the last line, we can simplify further as

$$\int \ln \left(\frac{p(\mathbf{x}, \mathbf{y}; \boldsymbol{\theta})}{p(\mathbf{y}|\mathbf{x}; \hat{\boldsymbol{\theta}})} \right) p(\mathbf{y}|\mathbf{x}; \hat{\boldsymbol{\theta}}) dy = \int \ln p(\mathbf{x}, \mathbf{y}; \boldsymbol{\theta}) p(\mathbf{y}|\mathbf{x}; \hat{\boldsymbol{\theta}}) dy \tag{2.96}$$

$$- \int \ln p(\mathbf{y}|\mathbf{x}; \hat{\boldsymbol{\theta}}) p(\mathbf{y}|\mathbf{x}; \hat{\boldsymbol{\theta}}) dy, \tag{2.97}$$

from which we define two quantities, namely

$$U(\boldsymbol{\theta}, \hat{\boldsymbol{\theta}}) = \int \ln p(\mathbf{y}, \mathbf{x}; \boldsymbol{\theta}) p(\mathbf{y}|\mathbf{x}; \hat{\boldsymbol{\theta}}) dy \tag{2.98}$$

$$= E_{\mathbf{y}} \left\{ \ln p(\mathbf{y}, \mathbf{x}; \boldsymbol{\theta}) \middle| \mathbf{x}; \hat{\boldsymbol{\theta}} \right\} \tag{2.99}$$

and

$$V(\boldsymbol{\theta}, \hat{\boldsymbol{\theta}}) = \int \ln p(\mathbf{y}; \boldsymbol{\theta}) p(\mathbf{y}|\mathbf{x}; \hat{\boldsymbol{\theta}}) d\mathbf{y} \tag{2.100}$$

$$= \mathrm{E}_{\mathbf{y}} \left\{ \ln p(\mathbf{y}; \boldsymbol{\theta}) \Big| \mathbf{x}; \hat{\boldsymbol{\theta}} \right\}. \tag{2.101}$$

From these definitions and (2.94), it follows that

$$\mathcal{L}(\boldsymbol{\theta}) \geq U(\boldsymbol{\theta}, \hat{\boldsymbol{\theta}}) - V(\hat{\boldsymbol{\theta}}, \hat{\boldsymbol{\theta}}). \tag{2.102}$$

Assuming continuity of the functions, we can choose a $\boldsymbol{\theta}^\star$ such that $U(\boldsymbol{\theta}^\star, \hat{\boldsymbol{\theta}}) \geq U(\hat{\boldsymbol{\theta}}, \hat{\boldsymbol{\theta}})$, implying the following relation between the log-likelihood of $\hat{\boldsymbol{\theta}}$ and $\boldsymbol{\theta}^\star$:

$$\mathcal{L}(\boldsymbol{\theta}^\star) \geq U(\boldsymbol{\theta}^\star, \hat{\boldsymbol{\theta}}) - V(\hat{\boldsymbol{\theta}}, \hat{\boldsymbol{\theta}}) \tag{2.103}$$

$$\geq U(\hat{\boldsymbol{\theta}}, \hat{\boldsymbol{\theta}}) - V(\hat{\boldsymbol{\theta}}, \hat{\boldsymbol{\theta}}) \tag{2.104}$$

$$= \mathcal{L}(\hat{\boldsymbol{\theta}}). \tag{2.105}$$

This means that by maximizing $U(\boldsymbol{\theta}, \hat{\boldsymbol{\theta}})$ we also increase the log-likelihood of the observed data. This is the fundamental idea of the EM algorithm.

We will now proceed to derive the specifics of the multi-pitch estimation problem based on the parametric signal model. First, we write the observed signal model in (1.5) as a sum of K sources in white additive Gaussian noise, i.e., $\mathbf{x} = \sum_{k=1}^{K} \mathbf{x}_k \in \mathbb{C}^N$, where the individual sources are given by

$$\mathbf{x}_k = \mathbf{Z}_k \mathbf{a}_k + \beta_k \mathbf{e}. \tag{2.106}$$

Here, the noise source is arbitrarily decomposed into K sources as $\mathbf{e}_k(n) = \beta_k \mathbf{e}$ where $\beta_k \geq 0$ is chosen so that $\sum_{k=1}^{K} \beta_k = 1$. In the framework of the EM algorithm, the set of vectors $\{\mathbf{x}_k\}$ is referred to as the complete data, which is unobservable, while the observed data is the vector $\mathbf{x}$. These vectors are assumed to be jointly Gaussian. The two data sets are related through a many-to-one mapping and this vector is therefore referred to as the incomplete data. The problem of interest is then to estimate the complete data set, or in our case its parameters, from the incomplete data. We start out by stacking our complete data in a vector $\mathbf{y}$ as

$$\mathbf{y} = \left[\mathbf{x}_1^T \ \mathbf{x}_2^T \ \cdots \ \mathbf{x}_K^T \right]^T. \tag{2.107}$$

We can now write the incomplete data as a function of this vector, i.e., $\mathbf{x} = \mathbf{H}\mathbf{y}$, where $\mathbf{H}$ is composed of K identity matrices having dimensions $N \times N$, i.e., $\mathbf{H} = \left[\mathbf{I} \cdots \mathbf{I} \right]$. Note that his mapping is completely arbitrary and that we could have picked others.

In each iteration, where (i) denotes the iteration number, the EM algorithm consists of two steps. The first, termed the expectation- or E-step, is the calculation of the conditional expectation of the log-likelihood function, i.e.,

$$U(\boldsymbol{\theta}, \boldsymbol{\theta}^{(i)}) = \int \ln p(\mathbf{y}, \mathbf{x}; \boldsymbol{\theta}) p(\mathbf{y}|\mathbf{x}; \boldsymbol{\theta}^{(i)}) d\mathbf{y}, \tag{2.108}$$

where $\boldsymbol{\theta}^{(i)}$ is a vector containing the ith iteration estimates of the parameters and $\boldsymbol{\theta}$ is the unknown parameter vector that parameterizes the likelihood function. Note that the function in (2.108) may be considered a function only of the complete data, since the observed data is completely specified by the complete data. Using (2.108), updated parameters are found in the maximization- or M-step by maximizing the above function, i.e.,

$$\boldsymbol{\theta}^{(i+1)} = \arg \max_{\boldsymbol{\theta}} U(\boldsymbol{\theta}, \boldsymbol{\theta}^{(i)}). \tag{2.109}$$

The two steps of the EM algorithm are particularly simple due to the noise term being Gaussian and white (for additional details, we refer to [51] and the references therein). The E-step reduces to the following where an estimate of the kth source in noise is obtained based on the parameters of the previous iteration:

$$\hat{\mathbf{x}}_k^{(i)} = \mathbf{Z}_k^{(i)} \hat{\mathbf{a}}_k^{(i)} + \beta_k \left(\mathbf{x} - \sum_{k=1}^{K} \mathbf{Z}_k^{(i)} \hat{\mathbf{a}}_k^{(i)} \right), \tag{2.110}$$

where $\mathbf{Z}_k^{(i)}$ is the Vandermonde matrix constructed from the fundamental frequency estimate $\hat{\omega}_k^{(i)}$. The problem of estimating the fundamental frequencies then becomes

$$\hat{\omega}_k^{(i+1)} = \arg \max_{\omega_k} \hat{\mathbf{x}}_k^{(i)H} \mathbf{Z}_k \left(\mathbf{Z}_k^H \mathbf{Z}_k \right)^{-1} \mathbf{Z}_k^H \hat{\mathbf{x}}_k^{(i)} \tag{2.111}$$

$$\approx \arg \max_{\omega_k} \hat{\mathbf{x}}_k^{(i)H} \mathbf{Z}_k \mathbf{Z}_k^H \hat{\mathbf{x}}_k^{(i)} \tag{2.112}$$

$$= \arg \max_{\omega_k} \sum_{l=1}^{L_k} |\hat{X}_k^{(i)}(\omega_k l)|^2, \tag{2.113}$$

where $\hat{X}_k^{(i)}(\omega)$ is the Fourier transform of the source estimate, i.e., $\hat{x}_k^{(i)}(n)$. The amplitudes are needed to form the source estimates in (2.110) and can be found, given $\hat{\omega}_k^{(i+1)}$, as

$$\hat{\mathbf{a}}_k^{(i+1)} = \left(\mathbf{Z}_k^{(i+1)H} \mathbf{Z}_k^{(i+1)} \right)^{-1} \mathbf{Z}_k^{(i+1)H} \hat{\mathbf{x}}_k^{(i)}, \tag{2.114}$$

or using the asymptotic approximation $\hat{\mathbf{a}}_k^{(i+1)} \approx \frac{1}{N} \mathbf{Z}_k^{(i+1)H} \hat{\mathbf{x}}_k^{(i)}$. The E-step in (2.110) and the M-step (2.113) are then repeated until some convergence criterion is met. The M-step in (2.113) can be seen to be identical to the NLS, except that (2.113) operates on the estimated source vector $\hat{\mathbf{x}}_k^{(i)}$ rather than the observed signal vector $\mathbf{x}(n)$. This has also led authors to use the EM algorithm to separate multiple speakers [18]. It is worth noting that refined estimates can be obtained in this framework using a numerical optimization approach, like Newton's method, if desired.

As can be seen, the EM algorithm splits the difficult joint estimation problem into a number of much simpler estimation problems by estimating the individual sources. In each iteration of the algorithm, the log-likelihood of the observed data is increased and the algorithm is guaranteed to converge, at least to a local maximum, under mild conditions. The main difficulty in using the EM algorithm is in obtaining the initial parameter estimates required to estimate the individual sources in (2.110). One possibility is to use the fast multi-pitch estimation method discussed in Section 2.7 or the method considered next.

2.9 ANOTHER RELATED METHOD

Many different methods for pitch estimation have been proposed in very different contexts and many are based on the nonlinear least-squares method. One such method that seems to have received some attention stems from the research community devoted to wavelets and sparse decompositions (also known as atomic decompositions). It is known as the Harmonic Matching Pursuit [66]. It is based on the following idea: define the residual at iteration i as

$$r^{(i)}(n) = x(n) - \sum_{k=1}^{i} \sum_{l=1}^{L_k} a_{k,l} e^{j\omega_k l n},$$ (2.115)

which can also be written as

$$r^{(i)}(n) = r^{(i-1)}(n) - \sum_{l=1}^{L_i} a_{i,l} e^{j\omega_i l n}.$$ (2.116)

The idea is then to use this residual for estimating the model parameters iteratively for each source. The residual can be thought of as a noise estimate, i.e., an estimate of $e(n)$, that gets progressively better as the number of iterations grows. It is initialized as $r^{(0)}(n) = x(n)$, i.e., we start out with simply using the observed signal. The signal model is then built iteratively from this residual by estimating the model parameters by solving the following problem:

$$\min_{\omega_i, \mathbf{a}_i} \left\| \mathbf{r}^{(i-1)} - \mathbf{Z}_i \mathbf{a}_i \right\|_2^2,$$ (2.117)

where $\mathbf{r}^{(i)}$ is a vector containing the residual. Solving first for the complex amplitudes $\mathbf{a}_i$, we get

$$\hat{\mathbf{a}}_i = \left(\mathbf{Z}_i^H \mathbf{Z}_i \right)^{-1} \mathbf{Z}_i^H \mathbf{r}^{(i-1)}.$$ (2.118)

Substituting this into (2.117), we get the fundamental frequency estimator

$$\hat{\omega}_i = \arg\min_{\omega_i} \left\| \mathbf{r}^{(i-1)} - \mathbf{Z}_i \left(\mathbf{Z}_i^H \mathbf{Z}_i \right)^{-1} \mathbf{Z}_i^H \mathbf{r}^{(i-1)} \right\|_2^2$$ (2.119)

$$= \arg\max_{\omega_i} \mathbf{r}^{(i-1)H} \mathbf{Z}_i \left(\mathbf{Z}_i^H \mathbf{Z}_i \right)^{-1} \mathbf{Z}_i^H \mathbf{r}^{(i-1)}$$ (2.120)

$$\approx \arg\max_{\omega_i} \mathbf{r}^{(i-1)H} \mathbf{Z}_i \mathbf{Z}_i^H \mathbf{r}^{(i-1)},$$ (2.121)

where the last line is only a computationally efficient approximation. The model parameters are then obtained for $i = 1, \ldots, K$ by first generating the residual using (2.116), then finding the fundamental frequency that minimizes the 2-norm of the modeling error using (2.120) and the amplitudes using (2.118). It is a simple method that overcomes some, but not all, of the shortcomings of the fast multi-pitch estimation scheme in Section 2.7. It has the desirable property that the 2-norm, or whatever norm is used, decreases in each iteration, i.e., we are certain that in terms of modeling the observed signal, we are getting a better model. However, this is not to be confused with obtaining good estimates of the underlying parameters! Comparing the Harmonic Matching Pursuit to the EM algorithm in Section 2.8, we see that these methods are similar in structure; in fact, it is easy to see that the former algorithm is just a cruder version of the EM algorithm in which the parameters of early iterations are not re-estimated during later iterations. In this sense, one might consider the Harmonic Matching Pursuit a poor man's EM algorithm (except for the use of $\{\beta_k\}$ in forming the source estimates). It will result in the maximum likelihood estimates under the same conditions as the fast multi-pitch estimator in Section 2.7, i.e., asymptotically in N for Gaussian noise. It should also be noted that the method, as presented in [66], does not address the very important related problem of model order estimation. This can be solved by combining it with the approach described in Section 2.6.

2.10 HARMONIC FITTING

An alternative way to estimate the fundamental frequency is to first find the frequencies of the unconstrained model using one of the many available methods for doing that and then find the fundamental frequency that best match those frequencies. We will refer to such methods as harmonic fitting methods. Such methods are based on a principle known as the EXIP (EXtended Invariance Principle) [163] (see also [166]) where first the parameters of an unstructured models are found and then fitted to a structured model. This methodology has been employed extensively in the literature on pitch estimation (although its theoretical foundations have usually not been explored in detail), and here we will present a representative method based on this principle. The method in question is the so-called Markov-like weighted least-squares (WLS) method of [107] (see also the related method [17]), and we will follow the derivation and definitions of that paper. Let $\{\psi_{k,l}\}$ be the set of unconstrained frequencies estimated from $\mathbf{x}_k(n)$ and let $\{A_{k,l}\}$ and $\{\phi_{k,l}\}$ be the respective amplitudes and phases of the complex sinusoids having those frequencies. We will here assume that L_k is known and that the noise $e_k(n)$ is white Gaussian having variance σ_k^2. Define

$$\boldsymbol{\theta}_k' = \begin{bmatrix} A_{k,1} \; \phi_{k,1} \; \psi_{k,1} \; \cdots \; A_{k,L_k} \; \phi_{k,L_k} \; \psi_{k,L_k} \end{bmatrix}^T \tag{2.122}$$

and

$$\boldsymbol{\eta}_k' = \begin{bmatrix} \omega_k \; A_{k,1} \; \phi_{k,1} \; \cdots \; A_{k,L_k} \; \phi_{k,L_k} \end{bmatrix}^T. \tag{2.123}$$

The basic idea of the method is that there exists a full rank so-called selection matrix $\mathbf{S}' \in \mathbb{Z}^{3L_k \times (2L_k+1)}$ that relates the vectors $\boldsymbol{\theta}_k'$ and $\boldsymbol{\eta}_k'$ as

$$\boldsymbol{\theta}_k' = \mathbf{S}' \boldsymbol{\eta}_k'. \tag{2.124}$$

We can now use this relation to find an estimate of η_k' using parameters estimates $\hat{\theta}_k'$ obtained using the unconstrained model as

$$\hat{\eta}_k' = \arg\min_{\eta_k'} \left\| \mathbf{W}'^{\frac{1}{2}} \left(\hat{\theta}_k' - \mathbf{S}'\eta_k' \right) \right\|_2^2, \tag{2.125}$$

where $\mathbf{W}'$ is a weighting matrix defined as

$$\mathbf{W}' = \mathbf{CRLB}^{-1}(\hat{\theta}_k'), \tag{2.126}$$

and $\mathbf{CRLB}^{-1}(\hat{\theta}_k')$ is the inverse of the CRLB matrix. This estimator is known as the Markov-like WLS estimator. The CRLB matrix for the unconstrained model is, fortunately, well-documented in the literature and is asymptotically simple. Specifically, the matrix $\mathbf{W}'$ becomes diagonal for large N, i.e.,

$$\mathbf{W}' = \begin{bmatrix} \mathbf{W}_1' & & \mathbf{0} \\ & \ddots & \\ \mathbf{0} & & \mathbf{W}_{L_k}' \end{bmatrix}. \tag{2.127}$$

where the individual sub-matrices contain the inverse of the CRLB matrix for the individual sinusoids of the unconstrained model (see, e.g., [88, 153]), i.e.,

$$\mathbf{W}_l' = \frac{1}{\sigma_k^2} \begin{bmatrix} 2N & 0 & 0 \\ 0 & 2N\hat{A}_{k,l}^2 & N^2\hat{A}_{k,l}^2 \\ 0 & N^2\hat{A}_{k,l}^2 & \frac{2}{3}N^3\hat{A}_{k,l}^2 \end{bmatrix}. \tag{2.128}$$

Since $\mathbf{W}_l'$ is also block diagonal, it follows that the weighting in (2.125) does not lead to refined estimates of the amplitude and it may therefore be left out of our estimation problem. Consequently, we define $\theta_k \in \mathbb{R}^{2L_k}$ and $\eta_k \in \mathbb{R}^{L_k+1}$ like θ_k' and η_k' but without the amplitudes. Now we may rewrite (2.124) as

$$\theta_k = \begin{bmatrix} 0 & 1 & 0 & \cdots & 0 \\ 1 & 0 & 0 & \cdots & 0 \\ 0 & 0 & 1 & \cdots & 0 \\ 2 & 0 & 0 & \cdots & 0 \\ & \vdots & & & \vdots \\ 0 & 0 & 0 & \cdots & 1 \\ L_k & 0 & 0 & \cdots & 0 \end{bmatrix} \eta_k \triangleq \mathbf{S}\eta_k. \tag{2.129}$$

As before, we can state our estimator as the minimizer of the norm of the error between the left and the right side of this expression, i.e.,

$$\hat{\eta}_k = \arg\min_{\eta_k} \left\| \mathbf{W}^{\frac{1}{2}} \left(\hat{\theta}_k - \mathbf{S}\eta_k \right) \right\|_2^2, \tag{2.130}$$

where $\hat{\boldsymbol{\theta}}_k$ is a set of initial estimates found based on the unconstrained model. In this case, $\mathbf{W}$ is a block diagonal matrix constructed from the sub-matrices $\mathbf{W}_l$ defined as

$$\mathbf{W}_l = \frac{1}{\sigma_k^2} \begin{bmatrix} 2N\hat{A}_{k,l}^2 & N^2\hat{A}_{k,l}^2 \\ N^2\hat{A}_{k,l}^2 & \frac{2}{3}N^3\hat{A}_{k,l}^2 \end{bmatrix}. \tag{2.131}$$

Using these definitions, we can now simplify the cost function associated with our estimation problem as follows:

$$\begin{aligned}
J &= \left\| \mathbf{W}^{\frac{1}{2}} \left(\hat{\boldsymbol{\theta}}_k - \mathbf{S}\boldsymbol{\eta}_k \right) \right\|_2^2 \tag{2.132} \\
&= \frac{1}{\sigma_k^2} \sum_{l=1}^{L_k} \left(\begin{bmatrix} \hat{\phi}_{k,l} \\ \hat{\psi}_{k,l} \end{bmatrix} - \begin{bmatrix} \phi_{k,l} \\ \omega_k l \end{bmatrix} \right)^T \begin{bmatrix} 2N\hat{A}_{k,l}^2 & N^2\hat{A}_{k,l}^2 \\ N^2\hat{A}_{k,l}^2 & \frac{2}{3}N^3\hat{A}_{k,l}^2 \end{bmatrix} \left(\begin{bmatrix} \hat{\phi}_{k,l} \\ \hat{\psi}_{k,l} \end{bmatrix} - \begin{bmatrix} \phi_{k,l} \\ \omega_k l \end{bmatrix} \right) \\
&= \frac{1}{\sigma_k^2} \sum_{l=1}^{L_k} \hat{A}_{k,l}^2 ([2N(\hat{\phi}_{k,l} - \phi_{k,l}) + N^2(\hat{\psi}_{k,l} - l\omega_k)](\hat{\phi}_{k,l} - \phi_{k,l}) \\
&\quad + \left[N^2(\hat{\phi}_{k,l} - \phi_{k,l}) + \frac{2}{3}N^3(\hat{\psi}_{k,l} - l\omega_k) \right] (\hat{\psi}_{k,l} - l\omega_k)). \tag{2.133}
\end{aligned}$$

Since we are only interested in the fundamental frequency, we will now proceed to substitute the phases $\{\phi_{k,l}\}$ by estimates obtained from a certain candidate frequency ω_k. We do this by differentiating the cost function above with respect to $\phi_{k,l}$ and equating the result with zero, whereby we obtain

$$\phi_{k,l} = \hat{\phi}_{k,l} + \frac{N}{2}(\hat{\psi}_{k,l} - l\omega_k) \tag{2.134}$$

for $l = 1, \ldots, L_k$. By substituting this expression into (2.133), we get

$$J = \frac{N^3}{6\sigma_k^2} \sum_{l=1}^{L_k} \hat{A}_{k,l}^2 \left(\hat{\psi}_{k,l} - l\omega_k \right)^2. \tag{2.135}$$

By differentiating this expression by ω_k and setting the resulting expression equal to zero, we obtain the following estimate of the fundamental frequency:

$$\hat{\omega}_k = \frac{\sum_{l=1}^{L_k} l\hat{A}_{k,l}^2 \hat{\psi}_{k,l}}{\sum_{l=1}^{L_k} l^2 \hat{A}_{k,l}^2}. \tag{2.136}$$

This estimate is essentially closed-form, but it does require that the amplitudes and an initial set of unconstrained frequencies are found. The method was demonstrated to have excellent statistical performance approaching the CRLB for large SNRs and number of observations N [107]. In fact, it is asymptotically equivalent to the maximum likelihood estimate even if the initial frequency estimates are not statistically efficient, but only consistent as would be the case if the frequencies are found using MUSIC [150], MODE [161, 162] or ESPRIT [147]. However, as was demonstrated in [23], the method is more sensitive to adverse conditions than methods that do not rely on intermediate estimates of the unconstrained frequencies. It is easy to imagine that spurious frequency estimates may cause large errors in estimates obtained using (2.136) and the same is likely to be the case for similar methods based on harmonic fitting. Also, when some of the harmonics may be missing, the optimization problem becomes complicated. It should be noted that the extension of this scheme to the multi-pitch case is nontrivial, since it would require that the unconstrained model is partitioned into sets of sinusoids belonging to different sources. For an overview of methods for finding the unconstrained frequencies, see, e.g., [88, 135, 156].

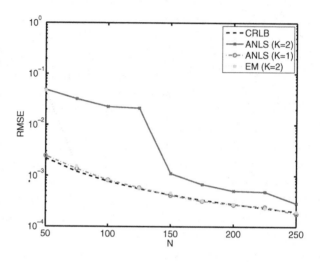

Figure 2.7: RMSE as a function of N with $PSNR = 10$ dB for one and two sources, respectively.

2.11   SOME RESULTS

In this section, we will report some basic experimental results regarding some of the statistical methods considered in this chapter. The results are obtained from a number of experiments based on Monte Carlo simulations using synthetic signals. In all experiments, signals were generated using the model in (1.5) with a white complex Gaussian noise source and 100 iterations were run for each free variable to estimate the RMSE. The experiments have been run for a single source having $\omega_1 = 0.2964$ and for two sources where an additional source having fundamental frequency $\omega_2 = 0.2257$. Both sources consist of three harmonics having unit amplitudes, i.e., $A_{k,l} = 1, \ \forall k, l$ and uniformly distributed phases that are randomized in each iteration. In the experiments, we show the performance of the methods of Sections 2.4, 2.7 and 2.8 as these are representative of common pitch estimators. Note that the methods of Sections 2.4 and 2.7 are denoted ANLS for approximate (or asymptotic) NLS in the figures. In Figure 2.7, the RMSE is shown as a function of the number of samples N for $PSNR = 10$ dB for one and two sources, respectively. Similarly, the RMSE is shown

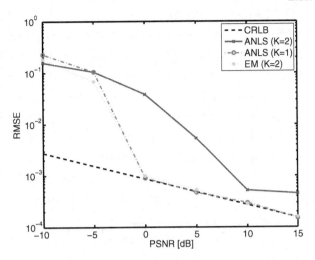

Figure 2.8: RMSE as a function of the PSNR with $N = 200$ for one and two sources, respectively.

as a function of the PSNR for $N = 200$ in Figure 2.8. Also shown in the figures, is the single-pitch CRLB, which is also a valid bound for the multi-pitch case, although an overly optimistic one when the harmonics of several sources overlap. Generally, one would expect the bound to be attainable when the harmonics of the sources are well-separated in frequency. The conclusions are clear: for the single-pitch case, the ANLS is actually very accurate, in fact, it attains the CRLB for the intervals of N and PSNRs tested here, so it can be claimed to be optimal under these conditions. However, when a second source is introduced, it does not perform well at all, as it ignores all interaction effects between the two sources. The EM algorithm, on the other hand, is able to mitigate this and attain the CRLB for reasonably high N and PSNRs. To investigate how the methods are able to handle closely spaced sources, the RMSE is shown as a function of the difference between the fundamental frequencies, i.e., $\Delta = |\omega_1 - \omega_2|$, in Figure 2.9 for $N = 250$ and $PSNR = 10$ dB. In this simulation, ω_1 is kept fixed at 0.2964 while ω_2 is varied from 0.2257 to 0.2964. It can be seen that the ANLS method does not perform well under these circumstances, while the EM algorithm, which is just an iterative NLS method, is, in fact, able to attain the CRLB until the two fundamental frequencies are too close. In Figure 2.10 and 2.11, the performance of the MAP-based order estimation criterion in (2.78) is shown in the form of the percentage of correctly estimated model orders as a function of N and the PSNR, respectively. The experimental conditions were as follows. A single-pitch signal having a fundamental frequency of 0.8170 and five unit amplitude harmonics in white Gaussian noise were used and 1000 Monte Carlo iterations were run. Note that the orders were found given the true fundamental frequency and that the noise variance was estimated using the exact ML formulation in (2.26). As can be seen from the two figures, the MAP criterion is able to find the correct model order almost all the time under these conditions.

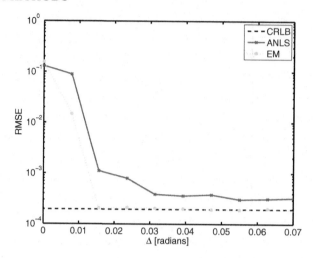

Figure 2.9: RMSE as a function of the difference between the fundamental frequencies of two source, i.e., $\Delta = |\omega_1 - \omega_2|$ for $N = 250$ and $PSNR = 10$ dB.

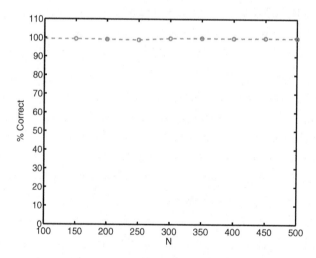

Figure 2.10: Percentage of correctly estimated model orders as a function of N with $PSNR = 10$ dB.

2.12   DISCUSSION

The methods considered in this chapter have a number of desirable features. They are statistically efficient when their exact formulations are used, i.e., without any use of approximations, meaning that they attain the CRLB for a high number of samples. For the single-pitch case, the associated optimization problem is one-dimensional nonlinear, something that, in principle, easily can be

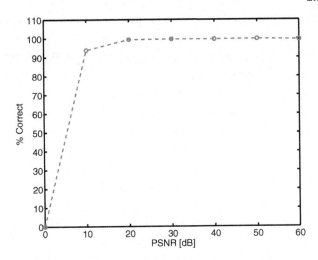

Figure 2.11: Percentage of correctly estimated model orders as a function of the PSNR with $N = 500$.

applied to real signals, although the complexity may be prohibitive for a large number of samples. It is, however, more problematic for the multi-pitch case where the associated problem is multi-dimensional. One can, of course, simply ignore that there are multiple sources, whereby one gets the method of Section 2.7. However, as shown in [33], the statistical performance of this method is, actually, quite poor, especially so for a low number of samples. This led to the use of the EM algorithm, which, as was also shown in [33], performs excellently, even under adverse conditions like when the fundamental frequencies of two sources are close. It does, however, suffer from two major drawbacks: it requires a lot of iterations for such cases and is thus computationally complex, and it has to be initialized appropriately or it will not converge to anything meaningful. It therefore offers only a partial solution–and this problem turns even worse when also the model and order selection problems are included. Another desirable property of the maximum likelihood methods is that simple implementations based on FFTs can be obtained for the problem at hand when the observation noise is white and Gaussian by asymptotic approximations that we know will get better as the number of samples increase. Also, a number of the problems associated with pitch estimation can be solved consistently within the framework of statistical estimation. This is, for example, the case of model selection and order estimation. It is also fairly easy to incorporate prior knowledge in the form of prior parameter distributions in the estimation process. When dealing with the estimation of sinusoidal frequencies, it is also well-known that even if the noise is colored, the LS-based methods are still asymptotically efficient, but the presence of colored noise may complicate the problems of model and order selection [30]. Given a set of unconstrained frequencies, it is also possible to combine these, in an optimal way, to form a fundamental frequency estimate, as we have seen in Section 2.10. In fact, this method has excellent statistical performance, as reported in [107]. It does, however, not perform well under adverse conditions (see [23]) as it turns out to be very sensitive to spurious frequency estimates. In the context of signal compression and modeling applications where the signal is reconstructed from the parameters, the methods based on the white Gaussian noise assumption also have the desirable property that they lead to a minimization of the reconstruction error in the sense of the 2-norm.

CHAPTER 3

Filtering Methods

3.1 INTRODUCTION

An intuitive idea for estimating the fundamental frequency is to filter the observed signal with a filter whose frequency response contains peaks at the harmonic frequencies. The fundamental frequency can then be found by maximizing the output power of the filter. Or, similarly, minimize the output power of a set of filters having notches at those frequencies. One could, for example, use a number of peak or notch filters for this, or perhaps design more flexible filters. Such a methodology has other applications too. It can be used for extracting the periodic contents of a signal from the background noise, i.e., signal enhancement or de-noising, or it can be used to extract one periodic source of many, i.e., signal separation.

In this chapter, we will pursue these ideas and present a number of different methods based on filtering. The first two methods are based on user designed filters, i.e., they are constrained to particular kinds of filters. Then, we consider more complicated designs based on signal adaptive optimal filters known from beamforming, i.e., spatial filtering and direction of arrival estimation.

The methods considered in this chapter can be applied both to single- and multi-pitch signals and have been demonstrated to have good statistical properties, i.e., low variance and low bias, for the multi pitch estimation problem, and we will, therefore, base our derivations on the signal $x(n)$, which may contain multiple periodic sources.

3.2 COMB FILTERING

That a signal is periodic means that some basic waveform repeats with a certain frequency. Mathematically, we may express this as $x(n) \approx x(n - D)$ where D is the repetition or pitch period. From this observation it follows that a measure of the periodicity can be obtained using a metric on $e(n)$ defined as

$$e(n) = x(n) - \alpha x(n - D), \tag{3.1}$$

where the constant $\alpha \in \mathbb{C}$ is introduced to allow for some variation of the amplitude. Taking the z-transform of this expression we get

$$E(z) = X(z) - \alpha X(z)z^{-D} \tag{3.2}$$
$$= X(z)(1 - \alpha z^{-D}), \tag{3.3}$$

which shows us that the process of matching a signal by a delayed version of itself can be seen as a filtering problem, where the output of the filter is the modeling error $e(n)$. More specifically, the transfer function $H(z)$ of the filter that operates on $x(n)$ can be seen to be

$$H(z) = \frac{E(z)}{X(z)} = (1 - \alpha z^{-D}). \tag{3.4}$$

This mathematical structure is known as a comb filter. The comb filtering approach has a rich history for fundamental frequency estimation and related problems, such as enhancement [109, 120, 122]. Usually, however, the comb filter in (3.4) is not used as it is restricted to integer pitch periods and is rather inefficient. Instead, one can derive more efficient methods based on notch filters. For a good introduction to notch and comb filters, we refer the interested reader to [126]. Notch filters are filters that cancel out, or attenuate, signal components at certain frequencies. In the complex case, these typically have the following form (see [122]):

$$H(z) = \frac{1 + \beta z^{-1}}{1 + \rho \beta z^{-1}} = \frac{P(z)}{P(\rho^{-1}z)}, \tag{3.5}$$

where β is a complex coefficient and $0 < \rho < 1$ is real. Since a periodic signal is comprised of possibly many harmonics, we can use L_k such notch filters having notches at frequencies $\{\psi_i\}$. Such a filter can be factorized into the following form

$$P(z) = \prod_{i=1}^{L_k} (1 - e^{j\psi_i} z^{-1}) \tag{3.6}$$

$$= 1 + \beta_1 z^{-1} + \beta_2 z^{-2} + \ldots + \beta_{L_k} z^{-L_k}, \tag{3.7}$$

which has zeros on the unit circle at the desired frequencies and complex coefficients $\{\beta_i\}$. For the real case, the notch frequencies would come in complex-conjugate pairs $\{\pm\psi_i\}$ and the corresponding coefficients would be real. The polynomial in (3.7) defines the numerator as in (3.5), while the denominator is

$$P(\rho^{-1}z) = \prod_{i=1}^{L_k} (1 - \rho e^{j\psi_i} z^{-1}) \tag{3.8}$$

$$= 1 + \rho \beta_1 z^{-1} + \rho^2 \beta_2 z^{-2} + \ldots + \rho^{L_k} \beta_{L_k} z^{-L_k}, \tag{3.9}$$

where $0 < \rho < 1$ is a parameter, which may be considered known or unknown, that leads to poles located inside the unit circle at the same angles as those of $P(z)$ but on a circle having a radius of ρ. Typical values for ρ are 0.95–0.995 [122]. In our case, the desired frequencies are given by $\psi_l = \omega_k l$, where ω_k is considered an unknown parameter, and the zeros of (3.7) are then located with regular intervals on the unit circle of the z-plane. In Figure 3.1, this is illustrated with the figure showing the poles and zeros of the comb filter in the z-plane. The frequency response of the same filter is plotted in Figure 3.2.

Note that in certain special cases, the polynomial defined in (3.7) reduces to a form similar to that of (3.4), but those special cases are generally not of interest here. Combining (3.7) and (3.9)

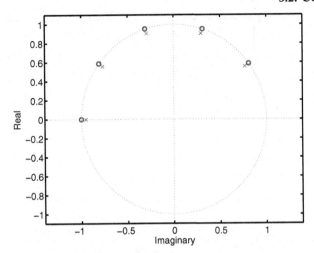

Figure 3.1: Z-plane representation of the zeros (circles) and poles (x-mark) of a comb filter for $\rho = 0.95$ and $\omega_k = 0.6283$ and $L_k = 5$.

in a manner similar to that of the notch filter in (3.5), we obtain the following filter:

$$H(z) = \frac{P(z)}{P(\rho^{-1}z)} = \frac{1 + \beta_1 z^{-1} + \beta_2 z^{-2} + \ldots + \beta_{L_k} z^{-L_k}}{1 + \rho\beta_1 z^{-1} + \rho^2\beta_2 z^{-2} + \ldots + \rho^{L_k}\beta_{L_k} z^{-L_k}}. \tag{3.10}$$

By applying this filter for various candidate fundamental frequencies to our observed signal $x(n)$, we can obtain the filtered signal $e(n)$ where the harmonics have been suppressed. This can be expressed as

$$E(z) = X(z)H(z), \tag{3.11}$$

which results in the following difference function from which the residual or error signal is readily obtained:

$$e(n) = x(n) + \beta_1 x(n-1) + \beta_2 x(n-2) + \ldots + \beta_{L_k} x(n-L_k)$$
$$- \rho\beta_1 e(n-1) - \rho^2\beta_2 e(n-2) - \ldots - \rho^{L_k}\beta_{L_k} e(n-L_k). \tag{3.12}$$

Finally, we can use this signal for defining the following metric:

$$J = \sum_{n=1}^{N} |e(n)|^2, \tag{3.13}$$

and this cost function can then be subject to optimization. The principle can be used for finding the fundamental frequency by calculating (3.13) as a function of the unknown fundamental frequency, i.e.,

$$\hat{\omega}_k = \arg \min_{\omega_k} J \tag{3.14}$$

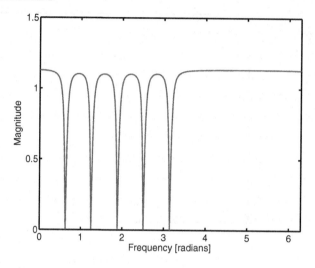

Figure 3.2: Frequency response (magnitude) of the comb filter for $\rho = 0.95$ and $\omega_k = 0.6283$ and $L_k = 5$.

and also ρ if desired. This is illustrated in Figure 3.3 where the cost function in (3.13) is shown for a signal having a fundamental frequency of 0.3142 and five harmonics. The fundamental idea of [122] is to use this principle in a recursive manner by minimizing (3.13) given an initial estimate of the fundamental frequency and the model order. Hereby, a computationally efficient scheme that can be used for either suppressing or enhancing the periodic contents of $x(n)$ is obtained. Note that in the scheme of [122], also the optimal ρ is found in the process. Interestingly, the comb filtering approach has been used in EM-like iterations (see Section 2.8) for multi-pitch estimation in [18].

3.3 FILTERBANK INTERPRETATION OF NLS

Before moving on to signal adaptive optimal filter methods, we will explore the relation between the maximum likelihood method for white Gaussian noise, i.e., the NLS method, and filtering methods. First, we construct a vector from M consecutive time-reversed samples of the observed signal, i.e.,

$$\mathbf{x}(n) = [\, x(n)\, x(n-1)\, \cdots\, x(n-M+1)\,]^T, \tag{3.15}$$

with $M \le N$. Next, we introduce the output signal $y_{k,l}(n)$ of the lth filter for the kth source having coefficients $h_{k,l}(n)$ as

$$y_{k,l}(n) = \sum_{m=0}^{M-1} h_{k,l}(m)x(n-m) = \mathbf{h}_{k,l}^H \mathbf{x}(n), \tag{3.16}$$

with $\mathbf{h}_{k,l}$ being a vector containing the filter coefficients of the lth filter, i.e.,

$$\mathbf{h}_{k,l} = [\, h_{k,l}(0)\, \cdots\, h_{k,l}(M-1)\,]^H. \tag{3.17}$$

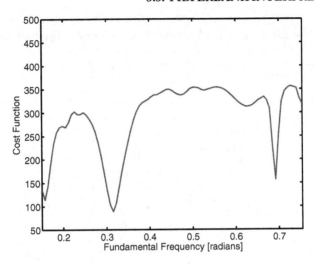

Figure 3.3: Cost function in (3.13) with $\rho = 0.95$ for a synthetic periodic signal. The estimate is obtained as the fundamental frequency that minimizes the cost function, i.e., the output power of the comb filter.

Defining the covariance matrix as $\mathbf{R} = \mathrm{E}\left\{\mathbf{x}(n)\mathbf{x}^H(n)\right\}$, the output power of the lth filter can be written as

$$\mathrm{E}\left\{|y_{k,l}(n)|^2\right\} = \mathrm{E}\left\{\mathbf{h}_{k,l}^H\mathbf{x}(n)\mathbf{x}^H(n)\mathbf{h}_{k,l}\right\} \tag{3.18}$$
$$= \mathbf{h}_{k,l}^H\mathbf{R}\mathbf{h}_{k,l}. \tag{3.19}$$

The total output power of all the filters is thus

$$\sum_{l=1}^{L_k}\mathrm{E}\left\{|y_{k,l}(n)|^2\right\} = \sum_{l=1}^{L_k}\mathbf{h}_{k,l}^H\mathbf{R}\mathbf{h}_{k,l}. \tag{3.20}$$

Defining a matrix $\mathbf{H}_k$ consisting of the filters $\{\mathbf{h}_{k,l}\}$ as

$$\mathbf{H}_k = \begin{bmatrix} \mathbf{h}_{k,1} & \cdots & \mathbf{h}_{k,L_k} \end{bmatrix}, \tag{3.21}$$

we can write the total output power as a sum of the powers of the indivudal subband signals, i.e.,

$$\sum_{l=1}^{L_k}\mathrm{E}\left\{|y_{k,l}(n)|^2\right\} = \mathrm{Tr}\left[\mathbf{H}_k^H\mathbf{R}\mathbf{H}_k\right]. \tag{3.22}$$

The task is then to design or choose a filter or a filterbank, and this can be done in several ways. Suppose we construct the filters from finite length complex sinusoids as

$$\mathbf{h}_{k,l} = \begin{bmatrix} e^{-j\omega_k l 0} & \cdots & e^{-j\omega_k l(M-1)} \end{bmatrix}^T, \tag{3.23}$$

which is the same as the vector $\mathbf{z}(\omega_k l)$ defined earlier except for the complex conjugation. The matrix $\mathbf{H}_k$ is, therefore, also identical to the Vandermonde matrix $\mathbf{Z}_k$, although we will here define it as[1]

$$\mathbf{Z}_k = [\ \mathbf{z}(\omega_k)\ \cdots\ \mathbf{z}(\omega_k L_k)\ ], \tag{3.24}$$

with $\mathbf{z}(\omega) = [\ 1\ e^{-j\omega}\ \cdots\ e^{-j\omega(M-1)}\ ]^T$. Then, we may write the total output power of the filterbank as

$$\mathrm{Tr}\left[\mathbf{H}_k^H \mathbf{R} \mathbf{H}_k\right] = \mathrm{Tr}\left[\mathbf{Z}_k^H \mathbf{R} \mathbf{Z}_k\right] \tag{3.25}$$

$$= \mathrm{E}\left\{\mathrm{Tr}\left[\mathbf{Z}_k^H \mathbf{x}(n)\mathbf{x}^H(n)\mathbf{Z}_k\right]\right\} \tag{3.26}$$

$$= \mathrm{E}\left\{\mathrm{Tr}\left[\mathbf{x}^H(n)\mathbf{Z}_k\mathbf{Z}_k^H \mathbf{x}(n)\right]\right\} \tag{3.27}$$

$$= \mathrm{E}\left\{\|\mathbf{Z}_k^H \mathbf{x}(n)\|_2^2\right\}. \tag{3.28}$$

By replacing the expectation operator by a finite sum over the realizations $\mathbf{x}(n)$, we get the FFT method introduced earlier, averaged over the sub-vectors $\mathbf{x}(n)$. Furthermore, by using only one sub-vector of length N, it becomes asymptotically equivalent (in N) to the nonlinear least-squares method and the white noise maximum likelihood method.

3.4   OPTIMAL FILTERBANK DESIGN

We will now look further into designing filters that are optimal given some criteria picked by the designer and given a sequence of the observed signal. The next couple of methods are based on equality constrained quadratic optimization, i.e., convex optimization. For an in-depth treatment of this subject, we refer to two excellent textbooks in the area, namely [14] and [6].

An intuitive approach is to find a set of filters that pass power undistorted at specific frequencies, in our case the harmonic frequencies, while minimizing the power at all other frequencies. This problem can be formulated mathematically as the optimization problem:

$$\min_{\mathbf{H}_k}\ \mathrm{Tr}\left[\mathbf{H}_k^H \mathbf{R} \mathbf{H}_k\right] \quad \text{s.t.} \quad \mathbf{H}_k^H \mathbf{Z}_k = \mathbf{I}, \tag{3.29}$$

where $\mathbf{I}$ is the $L_k \times L_k$ identity matrix. Furthermore, the matrix $\mathbf{Z}_k \in \mathbb{C}^{M \times L_k}$ has a Vandermonde structure and is constructed from L_k complex sinusoidal vectors as

$$\mathbf{Z}_k = [\ \mathbf{z}(\omega_k)\ \cdots\ \mathbf{z}(\omega_k L_k)\ ], \tag{3.30}$$

with $\mathbf{z}(\omega) = [\ 1\ e^{-j\omega}\ \cdots\ e^{-j\omega(M-1)}\ ]^T$. In words, the matrix contains the harmonically related complex sinusoids. The constraints basically state that the Fourier transform of the filters have unit gain at the harmonic frequencies. Using the method of Lagrange multipliers, the unconstrained

---

[1] That $\mathbf{Z}_k$ and $\mathbf{z}(\omega_k l)$ are the complex-conjugate of the usual definitions is due to the sub-vectors $\mathbf{x}(n)$ being time-reversed.

optimization problem can be written as

$$\mathcal{L}(\{\mathbf{h}_{k,l}\}, \{\boldsymbol{\lambda}_l\}) = \sum_{l=1}^{L_k} \mathbf{h}_{k,l}^H \mathbf{R} \mathbf{h}_{k,l} - \left(\mathbf{h}_{k,l}^H \mathbf{Z}_k - \mathbf{b}_l^T\right)\boldsymbol{\lambda}_l \qquad (3.31)$$

with $[\mathbf{b}_l]_v = 0$ for $v \neq l$ and $[\mathbf{b}_l]_v = 1$ for $v = l$, i.e.,

$$\mathbf{b}_l = [\, \underbrace{0 \,\cdots\, 0}_{l-1} \, 1 \, \underbrace{0 \,\cdots\, 0}_{L_k-l} \,] \qquad (3.32)$$

meaning that each individual filter is constrained to have unit gain for a certain harmonic frequency and zero gain for the others. It is easy to see that this can be written using a more convenient form as

$$\mathcal{L}(\mathbf{H}_k, \boldsymbol{\Lambda}) = \mathrm{Tr}\left\{\mathbf{H}_k^H \mathbf{R} \mathbf{H}_k\right\} - \mathrm{Tr}\left\{\left(\mathbf{H}_k^H \mathbf{Z}_k - \mathbf{I}\right)\boldsymbol{\Lambda}\right\}, \qquad (3.33)$$

where the matrix $\boldsymbol{\Lambda}$ contains all the Lagrange multiplier vectors $\boldsymbol{\lambda}_l$ associated with the various filters of the filterbank, i.e.,

$$\boldsymbol{\Lambda} = [\, \boldsymbol{\lambda}_1 \,\cdots\, \boldsymbol{\lambda}_{L_K} \,]. \qquad (3.34)$$

By differentiation with respect to the filters and the Lagrange multipliers, we obtain that the gradient of the composite cost function is

$$\nabla \mathcal{L}(\mathbf{H}_k, \boldsymbol{\Lambda}) = \begin{bmatrix} \mathbf{R}\mathbf{H}_k - \mathbf{Z}_k\boldsymbol{\Lambda} \\ -\mathbf{Z}_k^H \mathbf{H}_k + \mathbf{I} \end{bmatrix} \qquad (3.35)$$

$$= \begin{bmatrix} \mathbf{R} & -\mathbf{Z}_k \\ -\mathbf{Z}_k^H & 0 \end{bmatrix}\begin{bmatrix} \mathbf{H}_k \\ \boldsymbol{\Lambda} \end{bmatrix} + \begin{bmatrix} 0 \\ \mathbf{I} \end{bmatrix}. \qquad (3.36)$$

By setting these matrix equations equal to zero, we readily obtain that the Lagrange multipliers that solve the original problem are

$$\boldsymbol{\Lambda} = \left(\mathbf{Z}_k^H \mathbf{R}^{-1}\mathbf{Z}_k\right)^{-1}, \qquad (3.37)$$

and that the optimal filterbank expressed in terms of the Lagrange multipliers is

$$\mathbf{H}_k = \mathbf{R}^{-1}\mathbf{Z}_k\boldsymbol{\Lambda}. \qquad (3.38)$$

By substituting the solution for the Lagrange multipliers, the filter bank matrix $\mathbf{H}_k$ solving (3.29) can be seen to be given by

$$\mathbf{H}_k = \mathbf{R}^{-1}\mathbf{Z}_k\left(\mathbf{Z}_k^H \mathbf{R}^{-1}\mathbf{Z}_k\right)^{-1}. \qquad (3.39)$$

This data and frequency dependent filter bank can then be used to estimate the fundamental frequency by treating the fundamental frequency as an unknown variable and maximizing the power of the filter's output. This approach leads us to the following estimator:

$$\hat{\omega}_k = \arg\max_{\omega_k} \mathrm{Tr}\left[\left(\mathbf{Z}_k^H \mathbf{R}^{-1}\mathbf{Z}_k\right)^{-1}\right]. \qquad (3.40)$$

This expression depends only on the covariance matrix and the Vandermonde matrix constructed for different candidate fundamental frequencies. By tracking the covariance matrix (or its inverse) over time, signal adaptive optimal filters can be found. Since the optimal filter design in (3.39) and the estimator in (3.40) involve the inverse covariance matrix, $\mathbf{R}$, this matrix is required to have rank M, which is fulfilled for $M < N/2 + 1$. A practical way of implementing the estimator in (3.40) is to first evaluate the cost function on a coarse grid and then use a numerical optimization method for finding high-resolution estimates (see [33]).

In Figure 3.4, the output power of the optimal filterbank is shown as a function of the fundamental frequency for the signals shown in Figures 1.1 and 1.4. It can be seen that the cost function contains extremely narrow peaks and that the maximization of this cost function leads to an accurate estimate of the fundamental frequency. The same thing is shown in Figure 3.5, only this time for the mixture of the two signals in noise. From the last figure, it can be seen that the peaks corresponding to the fundamental frequencies of the two sources stand out rather clearly, yet the cost function is still very complicated.

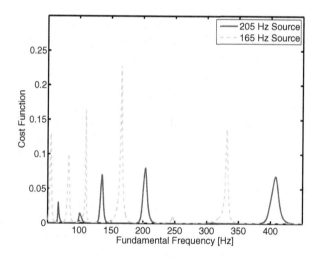

Figure 3.4: Cost function for the estimator based on the optimal filterbank obtain for the two signals in Figures 1.1 and 1.4, respectively.

3.5   OPTIMAL FILTER DESIGN

There is an alternative formulation of the filter design problem that we will now examine further. Suppose that we wish to design a single filter for the kth source, $\mathbf{h}_k$ that passes the signal undistorted at the harmonic frequencies and suppresses everything else. This filter design problem can be stated

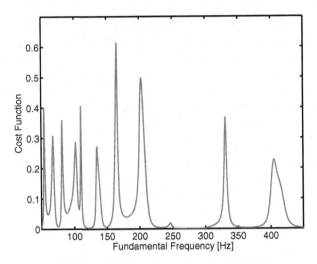

Figure 3.5: Cost function for the estimator based on the optimal filterbank for a mixture of the two signals in Figures 1.1 and 1.4 having fundamental frequencies 205 and 165 Hz, respectively.

mathematically as

$$\min_{\mathbf{h}_k} \mathbf{h}_k^H \mathbf{R} \mathbf{h}_k \quad \text{s.t.} \quad \mathbf{h}_k^H \mathbf{z}(\omega_k l) = 1, \tag{3.41}$$

$$\text{for} \quad l = 1, \ldots, L_k.$$

It is worth noting that the single filter in (3.41) is designed subject to L_k constraints, whereas in (3.29) the filter bank is designed using a number of constraints for each filter. Clearly, these two formulations are related; we will return to this relation later on. First, we will derive the optimal filter. Introducing the Lagrange multipliers $\boldsymbol{\lambda} = \begin{bmatrix} \lambda_1 & \cdots \lambda_{L_k} \end{bmatrix}^T$, the Lagrangian dual function associated with the problem stated above can be written as

$$\mathcal{L}(\mathbf{h_k}, \boldsymbol{\lambda}) = \mathbf{h}_k^H \mathbf{R} \mathbf{h}_k - \left(\mathbf{h}_k^H \mathbf{Z} - \mathbf{1}^T \right) \boldsymbol{\lambda} \tag{3.42}$$

with $\mathbf{1} = [\ 1 \ \cdots \ 1\ ]^T$. Taking the derivative with respect to the unknown filter impulse response, $\mathbf{h}_k$, and the Lagrange multipliers, we get

$$\nabla \mathcal{L}(\mathbf{h}_k, \boldsymbol{\lambda}) = \begin{bmatrix} \mathbf{R} \mathbf{h}_k - \mathbf{Z}_k \boldsymbol{\lambda} \\ -\mathbf{Z}_k^H \mathbf{h}_k + \mathbf{1} \end{bmatrix} \tag{3.43}$$

$$= \begin{bmatrix} \mathbf{R} & -\mathbf{Z}_k \\ -\mathbf{Z}_k^H & \mathbf{0} \end{bmatrix} \begin{bmatrix} \mathbf{h}_k \\ \boldsymbol{\lambda} \end{bmatrix} + \begin{bmatrix} \mathbf{0} \\ \mathbf{1} \end{bmatrix}. \tag{3.44}$$

By setting this expression equal to zero, i.e., $\nabla \mathcal{L}(\mathbf{h_k}, \boldsymbol{\lambda}) = \mathbf{0}$, and solving for the unknowns, we obtain, as with the filterbank design, the optimal Lagrange multipliers for which the equality constraints are satisfied as

$$\lambda = \left(\mathbf{Z}_k^H \mathbf{R}^{-1} \mathbf{Z}_k\right)^{-1} \mathbf{1} \tag{3.45}$$

and the optimal filter as

$$\mathbf{h}_k = \mathbf{R}^{-1} \mathbf{Z}_k \lambda. \tag{3.46}$$

By combining the last two expressions, we get the optimal filter expressed in terms of the covariance matrix and the Vandermonde matrix $\mathbf{Z}_k$, i.e.,

$$\mathbf{h}_k = \mathbf{R}^{-1} \mathbf{Z}_k \left(\mathbf{Z}_k^H \mathbf{R}^{-1} \mathbf{Z}_k\right)^{-1} \mathbf{1}. \tag{3.47}$$

The output power of this filter can then be expressed as

$$\mathbf{h}_k^H \mathbf{R} \mathbf{h}_k = \mathbf{1}^H \left(\mathbf{Z}_k^H \mathbf{R}^{-1} \mathbf{Z}_k\right)^{-1} \mathbf{1}, \tag{3.48}$$

which, as for the first design, depends only on the inverse of $\mathbf{R}$ and the Vandermonde matrix $\mathbf{Z}_k$. By maximizing the output power, we readily obtain an estimate of the fundamental frequency as

$$\hat{\omega}_k = \arg\max_{\omega_k} \mathbf{1}^H \left(\mathbf{Z}_k^H \mathbf{R}^{-1} \mathbf{Z}_k\right)^{-1} \mathbf{1}. \tag{3.49}$$

In Figure 3.6, the frequency response of the optimal filter in (3.48) is depicted for $\omega_k = 0.3142$ and $L_k = 5$ for white noise, i.e., $\mathbf{R} = \sigma^2 \mathbf{I}$, which leads to the optimal filter $\mathbf{h}_k = \mathbf{Z}_k \left(\mathbf{Z}_k^H \mathbf{Z}_k\right)^{-1} \mathbf{1}$. The filter length is $M = 50$.

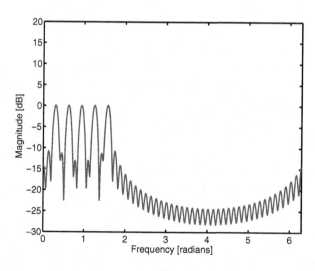

Figure 3.6: Frequency response (magnitude) of the optimal filter for $\omega_k = 0.3142$ and $L_k = 5$ for white noise.

3.6   ASYMPTOTIC ANALYSIS

We will now relate the two filter design methods and the associated estimators in (3.40) and (3.49). On the one hand, the optimization problem in (3.29) allows for more degrees of freedom, since L_k filters of length M are designed, while (3.41) involves only a single filter. On the other hand, the former design is based on L_k^2 constraints as opposed to the latter approach only involving L_k. Comparing the optimal filters in (3.39) and (3.47), we observe that the latter can be written in terms of the former as

$$\mathbf{h}_k = \mathbf{R}^{-1}\mathbf{Z}_k \left(\mathbf{Z}_k^H \mathbf{R}^{-1}\mathbf{Z}_k\right)^{-1} \mathbf{1} \tag{3.50}$$

$$= \mathbf{H}_k \mathbf{1} = \sum_{l=1}^{L} \mathbf{h}_{k,l}, \tag{3.51}$$

so, clearly, the two methods are related. Using this to rewrite the output power in (3.48), we get

$$\mathbf{h}_k^H \mathbf{R}\mathbf{h}_k = \left(\sum_{l=1}^{L} \mathbf{h}_{k,l}^H\right) \mathbf{R} \left(\sum_{m=1}^{L} \mathbf{h}_{k,m}\right), \tag{3.52}$$

as opposed to

$$\mathrm{Tr}\left[\mathbf{H}_k^H \mathbf{R}\mathbf{H}_k\right] = \sum_{l=1}^{L} \mathbf{h}_{k,l}^H \mathbf{R}\mathbf{h}_{k,l} \tag{3.53}$$

for the filterbank approach. It can be seen that the single-filter approach includes the cross-terms $\mathbf{h}_{k,l}^H \mathbf{R}\mathbf{h}_{k,m}$, for $l \neq m$, while these do not appear in (3.53). From this it follows that the cost functions are generally different, i.e.,

$$\mathbf{1}^H \left(\mathbf{Z}_k^H \mathbf{R}^{-1}\mathbf{Z}_k\right)^{-1} \mathbf{1} \neq \mathrm{Tr}\left[\left(\mathbf{Z}_k^H \mathbf{R}^{-1}\mathbf{Z}_k\right)^{-1}\right] \tag{3.54}$$

$$\mathbf{h}_k^H \mathbf{R}\mathbf{h}_k \neq \mathrm{Tr}\left[\mathbf{H}_k^H \mathbf{R}\mathbf{H}_k\right], \tag{3.55}$$

with equality if the cross-terms cancel out or are zero. This means that the two filters will result in different output powers and thus possibly different estimates. Next, we will analyze the asymptotic properties of the cost function

$$\lim_{M \to \infty} M\mathbf{1}^H \left(\mathbf{Z}_k^H \mathbf{R}^{-1}\mathbf{Z}_k\right)^{-1} \mathbf{1}, \tag{3.56}$$

where M has been introduced to ensure convergence. For simplicity, we will in the following derivations assume that the power spectral density of $x(n)$ is finite and non-zero. Although this is strictly speaking not the case for our signal model, the analysis will nonetheless provide some insights into the properties of the filtering methods. In the analysis, we will make use of the following result (see, e.g., [7])

$$\lim_{M\to\infty} (\mathbf{AB}) = \left(\lim_{M\to\infty} \mathbf{A} \right) \left(\lim_{M\to\infty} \mathbf{B} \right) \tag{3.57}$$

where it is assumed that the limits $\lim_{M\to\infty} \mathbf{A}$ and $\lim_{M\to\infty} \mathbf{B}$ exist for the individual elements of $\mathbf{A}$ and $\mathbf{B}$. Furthermore, it can be shown that

$$\left(\lim_{M\to\infty} \mathbf{A}^{-1} \right) = \left(\lim_{M\to\infty} \mathbf{A} \right)^{-1}. \tag{3.58}$$

Applying (3.57) and (3.58) to the cost function in (3.56), we get

$$\lim_{M\to\infty} M \mathbf{1}^H \left(\mathbf{Z}_k^H \mathbf{R}^{-1} \mathbf{Z}_k \right)^{-1} \mathbf{1} = \mathbf{1}^H \left(\lim_{M\to\infty} \frac{1}{M} \mathbf{Z}_k^H \mathbf{R}^{-1} \mathbf{Z}_k \right)^{-1} \mathbf{1}. \tag{3.59}$$

We are now left with the problem of determining the limit

$$\lim_{M\to\infty} \frac{1}{M} \mathbf{Z}_k^H \mathbf{R}^{-1} \mathbf{Z}_k. \tag{3.60}$$

In doing so, we will make use of the asymptotic equivalence of Toeplitz and circulant matrices. For a given Toeplitz matrix, here $\mathbf{R}$, we can construct an asymptotically equivalent circulant $M \times M$ matrix $\mathbf{C}$ in the sense that [65]

$$\lim_{M\to\infty} \frac{1}{\sqrt{M}} \|\mathbf{C} - \mathbf{R}\|_F = 0, \tag{3.61}$$

where $\| \cdot \|_F$ is the Frobenius norm and the limit is taken over the dimensions of $\mathbf{C}$ and $\mathbf{R}$. The conditions under which this was derived in [65] apply to the noise covariance matrix when the stochastic components are generated by a moving average or a stable auto-regressive process. More specifically, the auto-correlation sequence has to be absolutely summable. The result also applies to the deterministic signal components as $\mathbf{Z}_k \mathbf{P}_k \mathbf{Z}_k$ is asymptotically the EVD of the covariance matrix of $\mathbf{Z}_k \mathbf{a}_k$ (except for a scaling–see also Section 1.5) and circulant. A circulant matrix $\mathbf{C}$ has the eigenvalue decomposition $\mathbf{C} = \mathbf{U}\mathbf{\Gamma}\mathbf{U}^H$ where $\mathbf{U}$ is the Fourier matrix. Thus, the complex sinusoids in $\mathbf{Z}_k$ are asymptotically eigenvectors of $\mathbf{R}$. This allows us to determine the limit as (see [65, 69])

$$\lim_{M\to\infty} \frac{1}{M} \mathbf{Z}_k^H \mathbf{R} \mathbf{Z}_k = \begin{bmatrix} \Phi_x(\omega_k) & & \mathbf{0} \\ & \ddots & \\ \mathbf{0} & & \Phi_x(\omega_k L_k) \end{bmatrix}, \tag{3.62}$$

with $\Phi_x(\omega)$ being the power spectral density of $x(n)$. Similarly, an expression for the inverse of $\mathbf{R}$ can be obtained as $\mathbf{C}^{-1} = \mathbf{U}\mathbf{\Gamma}^{-1}\mathbf{U}^H$ (again, see [65] for details). We now arrive at the following (see also [69] and [155]):

$$\lim_{M \to \infty} \frac{1}{M} \mathbf{Z}_k^H \mathbf{R}^{-1} \mathbf{Z}_k = \begin{bmatrix} \Phi_x^{-1}(\omega_k) & & \mathbf{0} \\ & \ddots & \\ \mathbf{0} & & \Phi_x^{-1}(\omega_k L_k) \end{bmatrix}. \tag{3.63}$$

Asymptotically, (3.49) can, therefore, be written as

$$\lim_{M \to \infty} M \mathbf{1}^H \left(\mathbf{Z}_k^H \mathbf{R}^{-1} \mathbf{Z}_k \right)^{-1} \mathbf{1} = \sum_{l=1}^{L_k} \Phi_x(\omega_k l), \tag{3.64}$$

which is simply the sum over the power spectral density evaluated at the harmonic frequencies. Similar derivations for the filterbank formulation yield

$$\lim_{M \to \infty} M \operatorname{Tr} \left[\left(\mathbf{Z}_k^H \mathbf{R}^{-1} \mathbf{Z}_k \right)^{-1} \right] = \sum_{l=1}^{L_k} \Phi_x(\omega_k l), \tag{3.65}$$

which is the same as (3.64). Note that for a finite M, the above expression still involves only the diagonal terms because of the trace. However, the diagonal terms are then not the power spectral density $\Phi_x(\omega)$ evaluated at certain points. Thus, from the above analysis, we conclude that the two cost functions are different for finite M and may yield different estimates, but are asymptotically equivalent. This equivalence is essentially due to $(\mathbf{Z}_k^H \mathbf{R}^{-1} \mathbf{Z}_k)^{-1}$ being asymptotically diagonal. In [26], the two approaches were also reported to have similar performance although the output power estimates deviate. Another interesting consequence of the analysis in this section is that the methods based on optimal filtering yield results that are asymptotically equivalent to those of obtained using the NLS method as described in Section 3.3.

3.7 INVERSE COVARIANCE MATRIX

The two methods based on optimal filtering involve the inverse covariance matrix. We will now analyze the properties of the estimators further by finding a closed-form expression for the inverse of the covariance matrix based on the covariance matrix model introduced in Section 1.5. For the single-pitch case, the covariance matrix model is

$$\mathbf{R}_k = \operatorname{E}\left\{ \mathbf{x}_k(n) \mathbf{x}_k^H(n) \right\} \tag{3.66}$$
$$= \mathbf{Z}_k \mathbf{P}_k \mathbf{Z}_k^H + \mathbf{Q}_k, \tag{3.67}$$

and, for simplicity, we will use this model in the following. A variation of the matrix inversion lemma states that for matrices $\mathbf{A}, \mathbf{B}, \mathbf{C}, \mathbf{D}$

$$(\mathbf{A} + \mathbf{B}\mathbf{C}\mathbf{D})^{-1} = \mathbf{A}^{-1} - \mathbf{A}^{-1}\mathbf{B} \left(\mathbf{C}^{-1} + \mathbf{D}\mathbf{A}^{-1}\mathbf{B} \right)^{-1} \mathbf{D}\mathbf{A}^{-1} \tag{3.68}$$

provided that the respective matrix inverses exist. It can be seen that (3.68) is of the same form as (3.67). This provides us with a useful closed-form expression of the covariance matrix model, i.e.,

$$\mathbf{R}_k^{-1} = \left(\mathbf{Z}_k \mathbf{P}_k \mathbf{Z}_k^H + \mathbf{Q}_k \right)^{-1} \tag{3.69}$$

$$= \mathbf{Q}_k^{-1} - \mathbf{Q}_k^{-1} \mathbf{Z}_k \left(\mathbf{P}_k^{-1} + \mathbf{Z}_k^H \mathbf{Q}_k^{-1} \mathbf{Z}_k \right)^{-1} \mathbf{Z}_k^H \mathbf{Q}_k^{-1}. \tag{3.70}$$

Note that $\mathbf{P}_k^{-1}$ exists for a set of sinusoids having distinct frequencies and non-zero amplitudes and so does the inverse noise covariance matrix $\mathbf{Q}_k^{-1}$ as long as the noise has non-zero variance.

Now, to analyze (3.70) further, we evaluate the estimators for a candidate fundamental frequency generating a Vandermonde matrix $\widehat{\mathbf{Z}}_k$. The cost functions of the estimators based on optimal filtering then involve the following:

$$\widehat{\mathbf{Z}}_k^H \mathbf{R}_k^{-1} \widehat{\mathbf{Z}}_k = \widehat{\mathbf{Z}}_k^H \mathbf{Q}_k^{-1} \widehat{\mathbf{Z}}_k - \widehat{\mathbf{Z}}_k^H \mathbf{Q}_k^{-1} \mathbf{Z}_k \left(\mathbf{P}_k^{-1} + \mathbf{Z}_k^H \mathbf{Q}_k^{-1} \mathbf{Z}_k \right)^{-1} \mathbf{Z}_k^H \mathbf{Q}_k^{-1} \widehat{\mathbf{Z}}_k, \tag{3.71}$$

and we will normalize this matrix and analyze its behavior as M tends to infinity, i.e.,

$$\lim_{M \to \infty} \frac{1}{M} \widehat{\mathbf{Z}}_k^H \mathbf{R}_k^{-1} \widehat{\mathbf{Z}}_k = \lim_{M \to \infty} \frac{1}{M} \widehat{\mathbf{Z}}_k^H \mathbf{Q}_k^{-1} \widehat{\mathbf{Z}}_k$$

$$- \lim_{M \to \infty} \frac{1}{M} \widehat{\mathbf{Z}}_k^H \mathbf{Q}_k^{-1} \mathbf{Z}_k \left(\lim_{M \to \infty} \frac{1}{M} \mathbf{P}_k^{-1} + \lim_{M \to \infty} \frac{1}{M} \mathbf{Z}_k^H \mathbf{Q}_k^{-1} \mathbf{Z}_k \right)^{-1} \lim_{M \to \infty} \frac{1}{M} \mathbf{Z}_k^H \mathbf{Q}_k^{-1} \widehat{\mathbf{Z}}_k. \tag{3.72}$$

Noting that $\lim_{M \to \infty} \frac{1}{M} \mathbf{P}_k^{-1} = 0$, we obtain

$$\lim_{M \to \infty} \frac{1}{M} \widehat{\mathbf{Z}}_k^H \mathbf{R}_k^{-1} \widehat{\mathbf{Z}}_k = \lim_{M \to \infty} \frac{1}{M} \widehat{\mathbf{Z}}_k^H \mathbf{Q}_k^{-1} \widehat{\mathbf{Z}}_k$$

$$- \lim_{M \to \infty} \frac{1}{M} \widehat{\mathbf{Z}}_k^H \mathbf{Q}_k^{-1} \mathbf{Z}_k \left(\lim_{M \to \infty} \frac{1}{M} \mathbf{Z}_k^H \mathbf{Q}_k^{-1} \mathbf{Z}_k \right)^{-1} \lim_{M \to \infty} \frac{1}{M} \mathbf{Z}_k^H \mathbf{Q}_k^{-1} \widehat{\mathbf{Z}}_k. \tag{3.73}$$

Furthermore, by substituting $\widehat{\mathbf{Z}}_k$ by $\mathbf{Z}_k$ in (3.72), i.e., by evaluating the expression for the true fundamental frequency, we get

$$\lim_{M \to \infty} \frac{1}{M} \mathbf{Z}_k^H \mathbf{R}_k^{-1} \mathbf{Z}_k = \lim_{M \to \infty} \frac{1}{M} \mathbf{Z}_k^H \mathbf{Q}_k^{-1} \mathbf{Z}_k - \lim_{M \to \infty} \frac{1}{M} \mathbf{Z}_k^H \mathbf{Q}_k^{-1} \mathbf{Z}_k$$

$$\times \left(\lim_{M \to \infty} \frac{1}{M} \mathbf{Z}_k^H \mathbf{Q}_k^{-1} \mathbf{Z}_k \right)^{-1} \lim_{M \to \infty} \frac{1}{M} \mathbf{Z}_k^H \mathbf{Q}_k^{-1} \mathbf{Z}_k = \mathbf{0}, \tag{3.74}$$

which shows that the expression tends to the zero matrix as M approaches infinity when the expression is evaluated for the true fundamental frequency. The normalized cost functions of the two optimal filtering approaches in (3.64) and (3.65), therefore, can be though of as tending towards infinity. The results of [65, 155] can be applied directly to determine the limit $\lim_{M \to \infty} \frac{1}{M} \widehat{\mathbf{Z}}_k^H \mathbf{Q}_k^{-1} \widehat{\mathbf{Z}}_k$, which occurs in several of the equations above. This is because the autocorrelation sequence of the noise processes $e_k(n)$ can safely be assumed to be absolutely summable and have a smooth and non-zero power spectral density $\Phi_{e_k}(\omega)$ for common noise sources such as stable auto-regressive and moving average processes. Therefore, the limit is

$$\lim_{M\to\infty} \frac{1}{M} \hat{Z}_k^H Q_k^{-1} \hat{Z}_k = \begin{bmatrix} \Phi_{e_k}^{-1}(\hat{\omega}_k) & & 0 \\ & \ddots & \\ 0 & & \Phi_{e_k}^{-1}(\hat{\omega}_k L_k) \end{bmatrix}. \tag{3.75}$$

For the white noise case, the noise covariance matrix is diagonal, i.e., $Q_k = \sigma_k^2 I$. In this case, the inverse of the covariance matrix model is

$$R_k^{-1} = Q_k^{-1} - Q_k^{-1} Z_k \left(P_k^{-1} + Z_k^H Q_k^{-1} Z_k \right)^{-1} Z_k^H Q_k^{-1} \tag{3.76}$$

$$= \frac{1}{\sigma_k^2} I - \frac{1}{\sigma_k^2} I Z_k \left(P_k^{-1} + Z_k^H \frac{1}{\sigma_k^2} I Z_k \right)^{-1} Z_k^H \frac{1}{\sigma_k^2} I \tag{3.77}$$

$$= \frac{1}{\sigma_k^2} \left(I - Z_k \left(\sigma_k^2 P_k^{-1} + Z_k^H Z_k \right)^{-1} Z_k^H \right). \tag{3.78}$$

Next, we note that asymptotically, the complex sinusoids in the columns of Z_k are orthogonal, i.e.,

$$\lim_{M\to\infty} \frac{1}{M} Z_k^H Z_k = I. \tag{3.79}$$

For large M (and thus N), the inverse covariance matrix can be approximated as

$$R_k^{-1} \approx \frac{1}{\sigma_k^2} \left(I - Z_k \left(\sigma_k^2 P_k^{-1} + M I_k \right)^{-1} Z_k^H \right). \tag{3.80}$$

It can be observed that the remaining matrix inversion involves two diagonal matrices that can be rewritten as

$$\sigma_k^2 P_k^{-1} + M I_k = \text{diag}\left(\begin{bmatrix} \frac{\sigma_k^2}{A_{k,1}^2} + M & \cdots & \frac{\sigma_k^2}{A_{k,L_k}^2} + M \end{bmatrix} \right), \tag{3.81}$$

which leads to the inverse

$$\left(\sigma_k^2 P_k^{-1} + M I \right)^{-1} = \text{diag}\left(\begin{bmatrix} \frac{A_{k,1}^2}{\sigma_k^2 + M A_{k,1}^2} & \cdots & \frac{A_{k,L_k}^2}{\sigma_k^2 + M A_{k,L_k}^2} \end{bmatrix} \right) \tag{3.82}$$

$$\triangleq \Gamma_k. \tag{3.83}$$

Finally, we arrive at the following expression, which is an asymptotic approximation of the inverse of the matrix covariance model:

$$R_k^{-1} \approx \frac{1}{\sigma_k^2} \left(I - Z_k \Gamma_k Z_k^H \right). \tag{3.84}$$

Interestingly, it can be seen that the inverse exhibits a similar structure as the covariance matrix model.

3.8   VARIANCE AND ORDER ESTIMATION

We will now show how to use these filters for estimating the variance of the signal once the harmonics have been filtered out. This is useful for determining the number of harmonics using the methods considered in Section 2.6. We will do this based on the filterbank design. First, we define an estimate of the noise as

$$\hat{e}(n) = x(n) - y_k(n), \tag{3.85}$$

which we will refer to as the residual. Moreover, $y_k(n)$ is the sum of the input signal filtered by the filterbank, i.e.,

$$y_k(n) = \sum_{m=0}^{M-1} \sum_{l=1}^{L_k} h_{k,l}(m) x(n-m) \tag{3.86}$$

$$= \sum_{m=0}^{M-1} h_k(m) x(n-m), \tag{3.87}$$

where $h_k(m)$ is the sum over the impulse responses of the filters of the filterbank. From the relation between the single filter design and the filterbank design in (3.51), it is now clear that when used this way, the two approaches lead to the same output signal $y_k(n)$. This also offers some insights into the difference between the designs in (3.29) and (3.41). More specifically, the difference is in the way the output power is measured, where (3.29) is based on the assumption that the power is additive over the filters. We can now write the noise estimate as

$$\hat{e}(n) = x(n) - \sum_{m=0}^{M-1} h_k(m) x(n-m) \tag{3.88}$$

$$\triangleq \mathbf{g}_k^H \mathbf{x}(n), \tag{3.89}$$

where $\mathbf{g}_k = [\,(1 - h_k(0))\ -h_k(1)\ \cdots\ -h_k(M-1)\,]^H$ is the modified filter. From the noise estimate, we can now estimate the noise variance for the L_kth order model as

$$\hat{\sigma}^2(L_k) = \mathrm{E}\left\{|\hat{e}(n)|^2\right\} = \mathrm{E}\left\{\mathbf{g}_k^H \mathbf{x}(n) \mathbf{x}^H(n) \mathbf{g}_k\right\} \tag{3.90}$$

$$= \mathbf{g}_k^H \mathbf{R} \mathbf{g}_k. \tag{3.91}$$

This expression is, however, not very convenient for a number of reasons: A notable property of the estimator in (3.49) is that it does not require the calculation of the filter and that the output power expression in (3.48) is simpler than the expression for the optimal filter in (3.47). To use (3.91) directly, we would first have to calculate the optimal filter using (3.47), then calculate the modified filter $\mathbf{g}_k$, before evaluating (3.91). Instead, we simplify the evaluation of (3.91) by defining the modified filter as $\mathbf{g}_k = \mathbf{b}_1 - \mathbf{h}_k$ where, as defined earlier,

$$\mathbf{b}_1 = [\,1\ 0 \cdots 0\,]^H. \tag{3.92}$$

Next, we use this definition to rewrite the variance estimate as

$$\hat{\sigma}^2(L_k) = \mathbf{g}_k^H \mathbf{R} \mathbf{g}_k = (\mathbf{b}_1 - \mathbf{h}_k)^H \mathbf{R}(\mathbf{b}_1 - \mathbf{h}_k) \tag{3.93}$$
$$= \mathbf{b}_1^H \mathbf{R} \mathbf{b}_1 - \mathbf{b}_1^H \mathbf{R} \mathbf{h}_k - \mathbf{h}_k^H \mathbf{R} \mathbf{b}_1 + \mathbf{h}_k^H \mathbf{R} \mathbf{h}_k. \tag{3.94}$$

The first term can be identified to equal the variance of the observed signal $x(n)$, i.e., $\mathbf{b}_1^H \mathbf{R} \mathbf{b}_1 = \mathrm{E}\{|x(n)|^2\}$, and $\mathbf{h}_k^H \mathbf{R} \mathbf{h}_k$ we know from (3.48). Writing out the cross-terms $\mathbf{b}_1^H \mathbf{R} \mathbf{h}_k$ using (3.47) yields

$$\mathbf{b}_1^H \mathbf{R} \mathbf{h}_k = \mathbf{b}_1^H \mathbf{R} \mathbf{R}^{-1} \mathbf{Z}_k \left(\mathbf{Z}_k^H \mathbf{R}^{-1} \mathbf{Z}_k \right)^{-1} \mathbf{1} \tag{3.95}$$
$$= \mathbf{b}_1^H \mathbf{Z}_k \left(\mathbf{Z}_k^H \mathbf{R}^{-1} \mathbf{Z}_k \right)^{-1} \mathbf{1}. \tag{3.96}$$

Furthermore, it can easily be verified that $\mathbf{b}_1^H \mathbf{Z}_k = \mathbf{1}^H$, from which it can be concluded that

$$\mathbf{b}_1^H \mathbf{R} \mathbf{h}_k = \mathbf{1}^H \left(\mathbf{Z}_k^H \mathbf{R}^{-1} \mathbf{Z}_k \right)^{-1} \mathbf{1} \tag{3.97}$$
$$= \mathbf{h}_k^H \mathbf{R} \mathbf{h}_k. \tag{3.98}$$

Therefore, the variance estimate can be expressed as

$$\hat{\sigma}^2(L_k) = \hat{\sigma}^2(0) - \mathbf{1}^H \left(\mathbf{Z}_k^H \mathbf{R}^{-1} \mathbf{Z}_k \right)^{-1} \mathbf{1}, \tag{3.99}$$

where $\hat{\sigma}^2(0) = \mathrm{E}\{|x(n)|^2\}$. The variance estimate in (3.99) conveniently features the same expression as in the fundamental frequency estimation criterion in (3.49). This means that the same expression can be used for determining the model order and the fundamental frequency, i.e., the approach allows for joint estimation of the model order and the fundamental frequency. The variance estimate in (3.99) also shows that the same filter that maximizes the output power minimizes the variance of the residual. A more conventional variance estimate could be formed by first finding the frequency using, e.g., (3.49) and then finding the amplitudes of the signal model using least-squares to obtain a noise variance estimate. Since the discussed procedure uses the same information in finding the fundamental frequency and the noise variance, it is superior to the least-squares approach in terms of computational complexity. Note that for finite filter lengths, the output of the filters considered here are generally "power levels" and not power spectral densities (see [100]), which is consistent with our use of the filters for estimating the variance. Asymptotically, the filters do comprise power spectral density estimates [26].

By inserting (3.99) in (2.78), the model order can be determined using the MAP criterion for a given fundamental frequency. By combining the variance estimate in (3.99) with (2.78), we obtain the following fundamental frequency estimator for the case of unknown model orders (for $L_k > 0$):

$$\hat{\omega}_k = \arg \min_{\omega_k} \min_{L_k} N \ln \hat{\sigma}^2(L_k) + \frac{3}{2} \ln N + L_k \ln N \tag{3.100}$$
$$= \arg \min_{\omega_k} \min_{L_k} N \ln \left(\hat{\sigma}^2(0) - \mathbf{1}^H \left(\mathbf{Z}_k^H \mathbf{R}^{-1} \mathbf{Z}_k \right)^{-1} \mathbf{1} \right) + \frac{3}{2} \ln N + L_k \ln N, \tag{3.101}$$

where the model order is also estimated in the process. To determine whether any harmonics are present at all, the criterion in (2.79) can be used. In Figure 3.7, a typical example of the cost function in (3.101) is plotted along with the log-likelihood term $N \ln \sigma^2(L_k)$ for 100 samples of a signal consisting of five harmonics having a fundamental frequency of 0.6283 in white Gaussian noise. It can be seen that for $L_k > 5$, the log-likelihood term does not decrease significantly, and as a result, the correct model order can be determined using the MAP criterion.

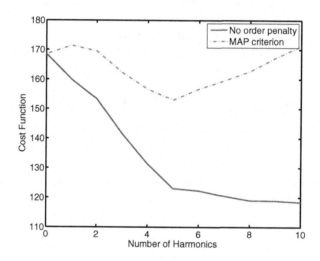

Figure 3.7: MAP model selection criterion and log-likelihood term for a periodic signal having five harmonics in white Gaussian noise.

3.9 FAST IMPLEMENTATION

Both the filterbank method and the single filter method require the calculation of the following matrix for every combination of candidate fundamental frequencies and orders:

$$\left(\mathbf{Z}_k^H \mathbf{R}^{-1} \mathbf{Z}_k \right)^{-1}, \tag{3.102}$$

where the respective cost function are formed as either the trace or the sum of all elements of this matrix. Since this requires a matrix inversion and matrix multiplications, all of cubic complexity, for each pair, there is a considerable computational burden of using these methods. We will now present an efficient implementation of the matrix inversion in (3.102). The inverse covariance matrix is less of a concern for two reasons. Firstly, it has to be calculated only once per segment, and secondly, many standard methods exist for updating the matrix inverse over time (see, e.g., [73]).

    The fast implementation that we will now derive, is based on the matrix inversion lemma, which states that for a matrix partitioned into matrices $\mathbf{A} \in \mathbb{C}^{m \times m}$, $\mathbf{B} \in \mathbb{C}^{n \times n}$, $\mathbf{C} \in \mathbb{C}^{m \times n}$, $\mathbf{D} \in$

$\mathbb{C}^{n \times m}$, the inverse can be calculated as:

$$\begin{bmatrix} \mathbf{A} & \mathbf{C} \\ \mathbf{D} & \mathbf{B} \end{bmatrix}^{-1} = \begin{bmatrix} \mathbf{A}^{-1} & \mathbf{O} \\ \mathbf{O} & \mathbf{O} \end{bmatrix} + \begin{bmatrix} -\mathbf{A}^{-1}\mathbf{C} \\ \mathbf{I} \end{bmatrix} \left(\mathbf{B} - \mathbf{D}\mathbf{A}^{-1}\mathbf{C} \right)^{-1} \begin{bmatrix} -\mathbf{D}\mathbf{A}^{-1} & \mathbf{I} \end{bmatrix}. \quad (3.103)$$

To apply this principle to the calculation of (3.102), we first define the matrix composed of vectors corresponding to the $L_k - 1$ first harmonics of the full matrix $\mathbf{Z}_k$ as

$$\mathbf{Z}_k^{(L_k-1)} = [\, \mathbf{z}(\omega_k) \, \cdots \, \mathbf{z}(\omega_k(L_k - 1)) \,], \quad (3.104)$$

and a vector containing the last harmonic L_k as

$$\mathbf{z}_k^{(L_k)} = \begin{bmatrix} e^{-j\omega_k L_k 0} & \cdots & e^{-j\omega_k L_k (M-1)} \end{bmatrix}^T. \quad (3.105)$$

Using these definitions, we can now rewrite (3.102) into a form similar to the left-hand side of (3.103), i.e.,

$$\left(\mathbf{Z}_k^H \mathbf{R}^{-1} \mathbf{Z}_k \right)^{-1} = \begin{bmatrix} \mathbf{Z}_k^{(L_k-1)H} \mathbf{R}^{-1} \mathbf{Z}_k^{(L_k-1)} & \mathbf{Z}_k^{(L_k-1)H} \mathbf{R}^{-1} \mathbf{z}_k^{(L_k)} \\ \mathbf{z}_k^{(L_k)H} \mathbf{R}^{-1} \mathbf{Z}_k^{(L_k-1)} & \mathbf{z}_k^{(L_k)H} \mathbf{R}^{-1} \mathbf{z}_k^{(L_k)} \end{bmatrix}^{-1} \quad (3.106)$$

$$\triangleq \boldsymbol{\Xi}_{L_k}, \quad (3.107)$$

where $\boldsymbol{\Xi}_{L_k}$ is the matrix (3.102) calculated for an order L_k model. Next, define the scalar quantity

$$\xi_{L_k} = \mathbf{z}_k^{(L_k)H} \mathbf{R}^{-1} \mathbf{z}_k^{(L_k)} \quad (3.108)$$

and the vector

$$\boldsymbol{\eta}_{L_k} = \mathbf{Z}_k^{(L_k-1)H} \mathbf{R}^{-1} \mathbf{z}_k^{(L_k)}. \quad (3.109)$$

We can now express the matrix in (3.102) in terms of the order $(L_k - 1)$ matrix $\boldsymbol{\Xi}_{L_k-1}, \xi_{L_k}$, and $\boldsymbol{\eta}_{L_k}$ using (3.103) as

$$\boldsymbol{\Xi}_{L_k} = \begin{bmatrix} \boldsymbol{\Xi}_{L_k-1} & \mathbf{0} \\ \mathbf{O} & 0 \end{bmatrix} + \begin{bmatrix} -\boldsymbol{\Xi}_{L_k-1}\boldsymbol{\eta}_{L_k} \\ 1 \end{bmatrix} \frac{1}{\xi_{L_k}^H - \boldsymbol{\eta}_{L_k}^H \boldsymbol{\Xi}_{L_k-1}\boldsymbol{\eta}_{L_k}} \begin{bmatrix} -\boldsymbol{\eta}_{L_k}^H \boldsymbol{\Xi}_{L_k-1} & 1 \end{bmatrix} \quad (3.110)$$

$$= \begin{bmatrix} \boldsymbol{\Xi}_{L_k-1} & \mathbf{0} \\ \mathbf{O} & 0 \end{bmatrix} + \frac{1}{\xi_{L_k}^H - \boldsymbol{\eta}_{L_k}^H \boldsymbol{\Xi}_{L_k-1}\boldsymbol{\eta}_{L_k}} \begin{bmatrix} \boldsymbol{\Xi}_{L_k-1}\boldsymbol{\eta}_{L_k}\boldsymbol{\eta}_{L_k}^H \boldsymbol{\Xi}_{L_k-1} & -\boldsymbol{\Xi}_{L_k-1}\boldsymbol{\eta}_{L_k} \\ -\boldsymbol{\eta}_{L_k}^H \boldsymbol{\Xi}_{L_k-1} & 1 \end{bmatrix}$$

$$(3.111)$$

$$\triangleq \begin{bmatrix} \boldsymbol{\Xi}_{L_k-1} & \mathbf{0} \\ \mathbf{O} & 0 \end{bmatrix} + \frac{1}{\beta_{L_k}} \begin{bmatrix} \varsigma_{L_k}\varsigma_{L_k}^H & -\varsigma_{L_k} \\ -\varsigma_{L_k}^H & 1 \end{bmatrix}. \quad (3.112)$$

This shows once $\boldsymbol{\Xi}_{L_k-1}$ is known, $\boldsymbol{\Xi}_{L_k}$ can be obtained in a simple way. To use this result to calculate the cost functions for the estimators (3.40) and (3.49) for a model order L_k, we proceed as follows. For a given ω_k, calculate the order 1 inverse matrix as

$$\boldsymbol{\Xi}_1 = \frac{1}{\xi_1}, \quad (3.113)$$

and then for $l = 2, \ldots, L_k$ calculate the required quantities and update the inverse matrix, i.e.,

$$\xi_l = \mathbf{z}_k^{(l)H} \mathbf{R}^{-1} \mathbf{z}_k^{(l)} \tag{3.114}$$

$$\eta_l = \mathbf{Z}_k^{(l-1)H} \mathbf{R}^{-1} \mathbf{z}_k^{(l)} \tag{3.115}$$

$$\zeta_l = \Xi_{l-1} \eta_l \tag{3.116}$$

$$\beta_l = \xi_l - \eta_l^H \Xi_{l-1} \eta_l \tag{3.117}$$

$$\Xi_l = \begin{bmatrix} \Xi_{l-1} & \mathbf{0} \\ \mathbf{O} & 0 \end{bmatrix} + \frac{1}{\beta_l} \begin{bmatrix} \zeta_l \zeta_l^H & -\zeta_l \\ -\zeta_l^H & 1 \end{bmatrix}, \tag{3.118}$$

using which the estimators in (3.40) and (3.49) along with the variance estimate in (3.99) can be implemented. We note in passing that, as usual, all the inner products involving complex sinusoids of different frequencies can be calculated efficiently using FFTs.

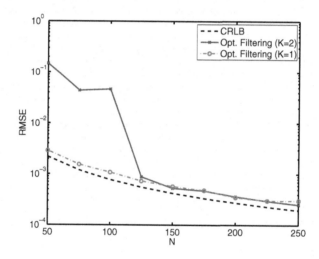

Figure 3.8: RMSE as a function of N with $PSNR = 10$ dB for one and two sources, respectively.

3.10   SOME RESULTS

We will end this chapter by first investigating the statistical properties of the optimal filtering approaches. As was shown in [27], the filterbank and single filter approaches have similar performance, although they appear to sometimes exhibit slightly different thresholding behaviour. For this reasons, we will evaluate the performance of the filterbank method in Monte Carlo simulations using synthetic signals generated from the ideal model in (1.5). The experiments are run for one and two sources having fundamental frequencies $\omega_1 = 0.2964$ and $\omega_2 = 0.2257$, three harmonics each and uniformly distributed phases and unit amplitudes, i.e., $A_{k,l} = 1$, $\forall\, k, l$. In each of the following experiments, 100 iterations were used to determine the RMSE for each point on the curves and a covariance matrix and filter length of $M = \lfloor N/5 \rfloor$ were used. First, the performance has been

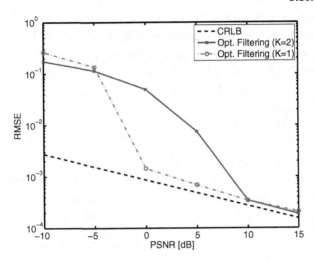

Figure 3.9: RMSE as a function of the PSNR with $N = 200$ for one and two sources, respectively.

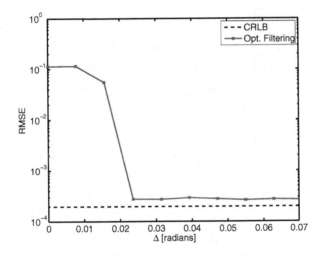

Figure 3.10: RMSE as a function of the difference between the fundamental frequencies of two source, i.e., $\Delta = |\omega_1 - \omega_2|$ for $N = 250$ and $PSNR = 10$ dB.

investigated as a function of N and the results are shown in Figure 3.8 for $PSNR = 10$ dB. It can be seen that the optimal filtering method under investigation (the filterbank method, to be specific) performs well for a single source, following but not attaining the CRLB. When a second source is introduced, some thresholding behavior can, though, be observed. The same general observations can be made of the RMSE as a function of the PSNR for $N = 200$ as seen in Figure 3.9; for one source, it performs well for the entire test range, while for two sources, the method exhibits some thresholding behavior. In Figure 3.10, the performance is shown as a function of $\Delta = |\omega_1 - \omega_2|$, i.e., the difference between the fundamental frequencies for a multi-pitch signal containing two sources with ω_1 being fixed at 0.2964 and ω_2 is varied from 0.2257 to 0.2964. The curve is obtain for $N = 250$ and $PSNR = 10$ dB. It can be seen that the method performs quite well, in fact, quite similarly to the ther other algorithms in terms of the thresholding behavior for closely spaced frequencies. As was demonstrated in Section 2.11, the MAP criterion works extremely well under the tested conditions, and we will now see what happens when the maximum likelihood noise variance estimate is replaced by one obtained using the optimal filtering approach presented in Section 3.8 by carrying out the same experiment, as before for a given ω_k. Note that, as we have seen, it does not matter, which of the two optimal designs is used for this application. 1000 Monte Carlo iterations were run for a single-pitch signal having fundamental frequency 0.8170, five harmonics and unit amplitudes. A filter length and covariance matrix of size $M = N/4$ was used in these experiments.

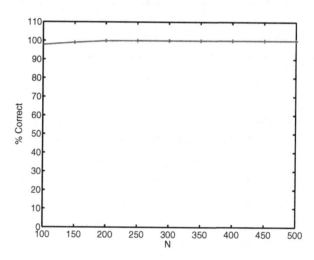

Figure 3.11: Percentage of correctly estimated model orders as a function of N with $PSNR = 10$ dB.

In Figure 3.11, the percentage of correctly estimated orders is shown as a function of the number of samples N for $PSNR = 40$ dB. Curiously, it can be observed that the method exhibits bad performance for low N. Since the only difference between this method and the one used in Section 3.8 is the noise variance estimate, we conclude that this estimate appears to be poor for low N. Note that this will also depends of the fundamental frequency. But for high N the method tends to estimating the true model order with a high accuracy. The method can be observed to perform well as a function of the PSNR in Figure 3.12 for a high number of samples, here $N = 500$, with the method breaking down when approaching 0 dB. Similarly, the importance of the choice of the filter length M has been investigated by keeping the number of observed samples N fixed at 200 samples and the PSNR at 40 dB while varying the M. The results are shown in Figure 3.13 and show that the method is indeed somewhat sensitive to the filter length, especially compared to other methods (see the respective

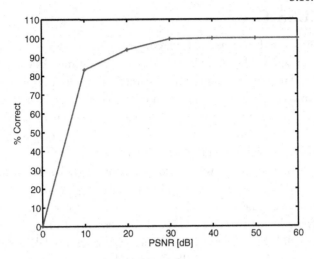

Figure 3.12: Percentage of correctly estimated model orders as a function of the PSNR with $N = 500$.

chapters). More specifically, it shows that M should not be chosen too close to $M/2$ or too low, and our prior choice of $M/4$ appears to be a good one.

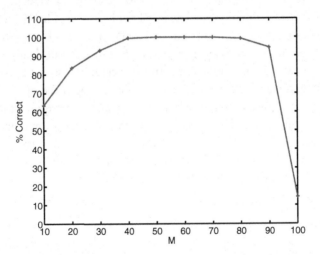

Figure 3.13: Percentage of correctly estimated model orders as a function of M for $N = 200$ and with $PSNR = 40$ dB.

3.11   DISCUSSION

The filtering approach to fundamental frequency estimation is very intuitive and it is, therefore, not surprising that such approaches have been used for a long time. As we have seen, the approximate maximum likelihood method for single-pitch estimation can also be interpreted as a filtering method. As a result, one would also expect these methods to have quite good statistical properties, and, as we have seen, this also turns out to be the case. The methods are generally not optimal in an estimation theoretical sense, i.e., they do not attain the CRLB, exhibiting a constant gap (in dB) as a function of N and the PSNR [33] (for more on the performance of the comb filtering approach for the single-pitch case, see [122]). They do, however, provide increasingly better estimates as a function of the number of samples and PSNR, as expected. For the multi-pitch case, the optimal filtering approaches have been shown to have some remarkable properties: it has excellent performance even under adverse conditions, for example, for two closely spaced sources [33]. In fact, it performs almost as well as the EM algorithm while being purely based on a single-pitch model, something that is not the case for neither the EM algorithm or the MUSIC method[2]. This can be explained by some observations that have been made in the application of similar optimal filters in array processing [106]. More specifically, it has been observed that the optimization procedure will result in filters that seek to suppress strong interfering sources, like other periodic sources. As we have seen, it is also possible to use the filtering approaches to obtain order estimates (and joint fundamental frequency and order estimates) by obtaining a variance estimate by filtering out the harmonics for different model orders and using statistical order estimation methods. However, it has been reported in [27] that this results in somewhat poor order estimates for a low number of samples. The major drawback of the optimal filtering approaches is the associated computational complexity. But as we have seen, it is possible to implement the methods recursively over the model order. Additionally, one can also use the FFT for calculating the cost function efficiently. In practice, it can sometimes also be quite difficult to locate the global extrema of the cost functions associated with the optimal filtering approaches as the peaks may be extremely narrow for large N and M. This requires that the cost functions are evaluated on a very fine grid and it may also be a challenge to perform numerical optimization using gradient or Hessian methods. Another drawback is that it is not always obvious how to incorporate prior knowledge in a consistent way, something that was easily done for the statistical methods.

---

[2]The MUSIC method depends on knowledge of the total number of sinusoids.

CHAPTER 4

Subspace Methods

4.1 INTRODUCTION

In subspace methods, the full space is divided into a subspace known as the signal subspace, that spans the space of the signal of interest, and its orthogonal complement, the so-called noise subspace. The properties of these subspaces are then exploited for various estimation and identification tasks (e.g., [71, 86, 98, 173, 174]). Subspace methods have a rich history in sinusoidal parameter estimation and are among the most eloquent estimators available today. Especially for the estimation of sinusoidal frequencies, a ubiquitous problem occurring in spectral estimation and direction-of-arrival problems in array processing, these methods have proven successful during the past three decades. Perhaps the most prominent subspace methods for frequency estimation are the Multiple Signal Classification (MUSIC) method [12, 150] and the Estimation of Signal Parameters by Rotational Invariance Techniques (ESPRIT) method of [147], while the earliest example of such methods is perhaps [128]. Fundamental frequency estimation, i.e., the problem of estimating the fundamental frequency of a set of harmonically related sinusoids, is an important component in many speech and audio processing systems. Until recently, however, this problem has not received much attention in the literature on subspace-based estimation. In [23], a joint fundamental frequency and order estimator based on the subspace orthogonality property of MUSIC was proposed and its application to analysis of speech and audio signals was demonstrated. In later publications, this method was extended to the multi-pitch case [21, 33] and the perturbed model [34]. In this chapter, we will present some subspace methods for fundamental frequency and order estimation along with the required background material.

4.2 SIGNAL AND NOISE SUBSPACE IDENTIFICATION

We start out this section, in which we will present some fundamental definitions, relations and results, by defining $\mathbf{x}(n)$ as a signal vector containing M samples of the observed signal, i.e., $\mathbf{x}(n) = [\, x(n)\, x(n+1)\, \cdots\, x(n+M-1)\,]^T$. Define the $M \times L$ Vandermonde matrix $\mathbf{Z}$ as consisting of the Vandermonde matrices for all the sources, i.e.,

$$\mathbf{Z} = [\, \mathbf{Z}_1 \; \cdots \; \mathbf{Z}_K \,], \tag{4.1}$$

where $L = \sum_{k=1}^{K} L_k$, and similarly a vector containing the complex amplitudes as

$$\mathbf{a}(n) = [\, \mathbf{a}_1^T(n) \; \cdots \; \mathbf{a}_K^T(n) \,]^T. \tag{4.2}$$

In the following, we assume that the all the frequencies are distinct. This is the case as long as the sources do not share any harmonics. We will start out by using our model of the observed signal, i.e.,

$$\mathbf{x}(n) = \sum_{k=1}^{K} \mathbf{Z}_k \mathbf{a}_k(n) + \mathbf{e}(n) \tag{4.3}$$

$$= \mathbf{Z}\mathbf{a}(n) + \mathbf{e}(n), \tag{4.4}$$

with $\mathbf{Z}_k \in \mathbb{C}^{M \times L_k}$ and $\mathbf{a}_k(n) \in \mathbb{C}^{L_k}$. Assuming that $\mathrm{E}\left\{\mathbf{e}(n)\mathbf{e}^H(n)\right\} = \sigma^2 \mathbf{I}$ and that the sources are statistically independent, the covariance matrix $\mathbf{R} \in \mathbb{C}^{M \times M}$ (with $L < M$) of the signal in (4.4) can be shown to be

$$\mathbf{R} = \mathrm{E}\left\{\mathbf{x}(n)\mathbf{x}^H(n)\right\} \tag{4.5}$$

$$= \sum_{k=1}^{K} \mathbf{Z}_k \mathbf{P}_k \mathbf{Z}_k^H + \sigma^2 \mathbf{I} \tag{4.6}$$

$$= \mathbf{Z}\mathbf{P}\mathbf{Z}^H + \sigma^2 \mathbf{I} \tag{4.7}$$

where $\mathrm{E}\left\{\mathbf{a}_k(n)\mathbf{a}_k^H(n)\right\} = \mathbf{P}_k$. It is worth noting that the decomposition in (4.6) does not depend on the noise being Gaussian, only on it being white. Assuming that the elements of $\mathbf{a}_k(n)$ are zero mean and statistically independent, the matrix $\mathbf{P}_k$ becomes diagonal, i.e.,

$$\mathbf{P}_k = \mathrm{diag}\left(\begin{bmatrix} A_{k,1}^2 & \cdots & A_{k,L_k}^2 \end{bmatrix}\right) \tag{4.8}$$

and

$$\mathbf{P} = \mathrm{diag}\left(\begin{bmatrix} \mathbf{P}_1 & \cdots & \mathbf{P}_K \end{bmatrix}\right). \tag{4.9}$$

Also, here σ^2 denotes the variance of the additive noise, $e(n)$, and $\mathbf{I}$ is the $M \times M$ identity matrix. It is important to realize that $\sum_{k=1}^{K} \mathbf{Z}_k \mathbf{P}_k \mathbf{Z}_k^H$ has rank $L = \sum_{k=1}^{K} L_k$, since all the columns of $\mathbf{Z}$ are linearly independent when the frequencies are distinct. Let

$$\mathbf{R} = \mathbf{U}\mathbf{\Lambda}\mathbf{U}^H \tag{4.10}$$

be the eigenvalue decomposition (EVD) of the covariance matrix. Since the covariance matrix is positive definite and symmetric by construction, $\mathbf{U}$ contains the M orthonormal eigenvectors of $\mathbf{R}$, i.e.,

$$\mathbf{U} = \begin{bmatrix} \mathbf{u}_1 & \cdots & \mathbf{u}_M \end{bmatrix}, \tag{4.11}$$

and $\mathbf{\Lambda}$ is a diagonal matrix containing the corresponding positive eigenvalues, λ_k, ordered as $\lambda_1 \geq \lambda_2 \geq \ldots \geq \lambda_M$. Additionally, it can be seen from (4.6) that $\lambda_{L+1} = \ldots = \lambda_M = \sigma^2$ due to the noise being white.

The subspace-based methods are based on a partitioning of the eigenvectors into a set belonging to the signal subspace $\mathcal{Z}$, spanned by the columns of $\mathbf{Z}$, and an orthogonal complement known as the noise subspace $\mathcal{G}$. Let $\mathbf{S}$ be formed from the eigenvectors corresponding to the L most significant eigenvalues, i.e.,

$$\mathbf{S} = \begin{bmatrix} \mathbf{u}_1 & \cdots & \mathbf{u}_L \end{bmatrix}. \tag{4.12}$$

We denote the subspace that is spanned by the columns of $\mathbf{S}$ $\mathcal{R}(\mathbf{S})$ and henceforth refer to it as the signal subspace. Similarly, let $\mathbf{G}$ be formed from the eigenvectors corresponding to the $M - L$ least significant eigenvalues, i.e.,

$$\mathbf{G} = \begin{bmatrix} \mathbf{u}_{L+1} & \cdots & \mathbf{u}_M \end{bmatrix}, \tag{4.13}$$

where $\mathcal{R}(\mathbf{G})$ is referred to as the noise subspace. Using these definitions, we can now manipulate the EVD in (4.10) to obtain $\mathbf{U}\left(\mathbf{\Lambda} - \sigma^2\mathbf{I}\right)\mathbf{U}^H = \sum_{k=1}^{K} \mathbf{Z}_k\mathbf{P}_k\mathbf{Z}_k^H$ and partition the EVD as

$$\mathbf{R} = \begin{bmatrix} \mathbf{S} & \mathbf{G} \end{bmatrix}\left(\begin{bmatrix} \mathbf{\Psi} & \mathbf{0} \\ \mathbf{0} & \mathbf{0} \end{bmatrix} + \sigma^2\mathbf{I}\right)\begin{bmatrix} \mathbf{S}^H \\ \mathbf{G}^H \end{bmatrix}, \tag{4.14}$$

where $\mathbf{\Psi} = \text{diag}\left(\begin{bmatrix} \lambda_1 - \sigma^2 & \cdots & \lambda_L - \sigma^2 \end{bmatrix}\right)$ is a diagonal matrix containing the eigenvalues of the deterministic parts of the covariance matrix. From this it is clear that we can express the covariance matrix of all the sinusoids as

$$\mathbf{S}\mathbf{\Psi}\mathbf{S}^H = \sum_{k=1}^{K} \mathbf{Z}_k\mathbf{P}_k\mathbf{Z}_k^H. \tag{4.15}$$

It must be stressed that the space spanned by the L most significant eigenvectors is the same space that is spanned by all the sinusoids and that a subset of either is not generally equivalent to a subset of the other.

4.3   SUBSPACE PROPERTIES

From equation (4.15), it is easy to see that the columns of $\mathbf{S}$, i.e., the signal subspace eigenvectors, span the same space as the columns of $\mathbf{Z}$ and that the columns of $\mathbf{Z}$, therefore, also are orthogonal to the columns of $\mathbf{G}$, i.e., the noise subspace eigenvectors. Mathematically, this can be written as

$$\mathbf{Z}^H\mathbf{G} = \mathbf{0} \tag{4.16}$$

and for sub-matrices corresponding to the individual sources as

$$\mathbf{Z}_k^H\mathbf{G} = \mathbf{0} \quad \forall k. \tag{4.17}$$

This is what we will refer to as the subspace orthogonality property. The idea of the MUSIC algorithm [12, 150] and the later refinement root-MUSIC [11] (see also [77]) is to exploit this relation for estimation purposes as follows: we can find an estimate of the matrix $\mathbf{G}$ from the EVD of the sample covariance matrix for an observed signal. Given this matrix, we can find the nonlinear parameters that characterize the columns of $\mathbf{Z}$ as the parameters that generate a $\mathbf{Z}$ that is orthogonal to $\mathbf{G}$. The remaining question is then how to measure orthogonality in a multi-dimensional space in a well-defined way. We will return to this question later on. The statistical properties of MUSIC for a known order have been studied extensively in [50, 157, 158].

There is a distinct property of the signal model considered in this book that can be used for estimation purposes and we will now explore this in more detail. By post-multiplication of (4.15) by $\mathbf{S}$, the following relation between the signal subspace eigenvectors and the Vandermonde matrix can be established (see [156]):

$$\mathbf{S} = \mathbf{ZB} \tag{4.18}$$

with $\mathbf{B} = \mathbf{PZ}^H\mathbf{S}\mathbf{\Psi}^{-1}$. Note that this expression can be used directly for finding parameter estimates (see, e.g., [98]). Next, we define the matrices $\underline{\mathbf{Z}}$ and $\overline{\mathbf{Z}}$, constructed by removing the last and first rows of $\mathbf{Z}$, i.e.,

$$\underline{\mathbf{Z}} = \begin{bmatrix} \mathbf{I} & \mathbf{0} \end{bmatrix} \mathbf{Z} \quad \text{and} \quad \overline{\mathbf{Z}} = \begin{bmatrix} \mathbf{0} & \mathbf{I} \end{bmatrix} \mathbf{Z}, \tag{4.19}$$

where $\mathbf{I}$ is now the $(M-1) \times (M-1)$ identity matrix. Similarly, we define from the signal subspace eigenvectors $\mathbf{S}$,

$$\underline{\mathbf{S}} = \begin{bmatrix} \mathbf{I} & \mathbf{0} \end{bmatrix} \mathbf{S} \quad \text{and} \quad \overline{\mathbf{S}} = \begin{bmatrix} \mathbf{0} & \mathbf{I} \end{bmatrix} \mathbf{S}. \tag{4.20}$$

From these definitions and (4.18), the matrices $\underline{\mathbf{S}}$ and $\underline{\mathbf{Z}}$ can be related through the matrix $\mathbf{B}$ as $\underline{\mathbf{S}} = \underline{\mathbf{Z}}\mathbf{B}$. Then, due to the particular structure of $\mathbf{Z}$, known as the shift-invariance property, the following can be seen to hold:

$$\overline{\mathbf{Z}} = \underline{\mathbf{Z}}\mathbf{D} \quad \text{and} \quad \overline{\mathbf{S}} = \underline{\mathbf{S}}\mathbf{\Xi}, \tag{4.21}$$

with $\mathbf{D} = \text{diag}\left(\begin{bmatrix} e^{j\psi_1} & \cdots & e^{j\psi_L} \end{bmatrix}\right)$. The matrix relating $\underline{\mathbf{S}}$ to $\overline{\mathbf{S}}$ can be written as

$$\mathbf{\Xi} = \mathbf{B}^{-1}\mathbf{DB}. \tag{4.22}$$

The matrices $\mathbf{\Xi}$ and $\mathbf{D}$ are thus related by a similarity transform. Note that it is not generally possible to relate the sub-matrices $\mathbf{Z}_k$ associated with the various sources to sub-matrices in $\mathbf{S}$ and the relation considered here thus applies only to the full matrix $\mathbf{Z}$.

4.4   PRE-WHITENING

Perhaps the biggest weakness of the subspace methods is that they rely on the noise being white. If the noise is not white, the decomposition of the covariance matrix and the partitioning of the eigenvectors into signal and noise subspaces may not hold, and, consequently, the signal and noise subspace eigenvectors cannot easily be identified from the eigenvalues. This is especially troublesome in connection with speech and audio signals where stochastic signal components are inherent parts of the signal and change all the time, often abruptly. In the array processing literature, where many of the subspace methods were first published, this problem is often suggested solved by pre-whitening of the signal, i.e., by applying a filter that makes the stochastic signal parts white. Consider the signal model in (4.4) for the case where the noise is not white. In that case, the covariance matrix model is

$$\mathbf{R} = E\left\{\mathbf{x}(n)\mathbf{x}^H(n)\right\} \tag{4.23}$$

$$= \mathbf{ZPZ}^H + \mathbf{Q}, \tag{4.24}$$

where $\mathbf{Q} = E\{\mathbf{e}(n)\mathbf{e}^H(n)\}$ is the noise covariance matrix. Let

$$\mathbf{Q} = \mathbf{V}\boldsymbol{\Gamma}\mathbf{V}^H \tag{4.25}$$

be the EVD of this matrix and define the matrix $\mathbf{W}$ as

$$\mathbf{W} = \boldsymbol{\Gamma}^{-\frac{1}{2}}\mathbf{V}^H. \tag{4.26}$$

By multiplying the observed signal vectors by this matrix, we get

$$E\left\{\mathbf{W}\mathbf{x}(n)\mathbf{x}^H(n)\mathbf{W}^H\right\} = \mathbf{W}E\left\{\mathbf{x}(n)\mathbf{x}^H(n)\right\}\mathbf{W}^H \tag{4.27}$$
$$= \mathbf{W}\mathbf{Z}\mathbf{P}\mathbf{Z}^H\mathbf{W}^H + \mathbf{W}\mathbf{Q}\mathbf{W}^H. \tag{4.28}$$

By substituting the noise covariance matrix by its EVD, we get

$$\mathbf{W}\mathbf{Q}\mathbf{W}^H = \mathbf{W}\mathbf{V}\boldsymbol{\Gamma}\mathbf{V}^H\mathbf{W}^H \tag{4.29}$$
$$= \boldsymbol{\Gamma}^{-\frac{1}{2}}\mathbf{V}^H\mathbf{V}\boldsymbol{\Gamma}\mathbf{V}^H\mathbf{V}\boldsymbol{\Gamma}^{-\frac{1}{2}} = \mathbf{I}. \tag{4.30}$$

So, by multiplying the input signal vector by the matrix $\mathbf{W}$, the noise covariance matrix is reduced to an identity matrix. One, therefore, refers to this matrix as a pre-whitening matrix. A computationally less demanding alternative to the pre-whitener in (4.26) is the Cholesky factor. Since covariance matrices are symmetric and positive definite, so are their inverses and the Cholesky factorization of the inverse noise covariance matrix $\mathbf{Q}^{-1}$ is (see, e.g., [64])

$$\mathbf{Q}^{-1} = \mathbf{L}\mathbf{L}^H, \tag{4.31}$$

where $\mathbf{L}$ is an $M \times M$ lower triangular matrix also known as the Cholesky factor. This matrix can also be used for pre-whitening by pre-multiplying $\mathbf{x}(n)$ by $\mathbf{L}^H$ as

$$\mathbf{L}^H\mathbf{Q}\mathbf{L} = \mathbf{L}^H\mathbf{L}^{-H}\mathbf{L}^{-1}\mathbf{L} = \mathbf{I}, \tag{4.32}$$

and this is preferable, since, once the pre-whitener has been calculated, the matrix-vector product $\mathbf{L}^H\mathbf{x}(n)$ can be more efficiently implemented than $\mathbf{W}\mathbf{x}(n)$ due to the triangular structure of $\mathbf{L}$. For the special case of pre-whitening for rank-deficient covariance matrices, we refer the reader to [72].

The principles discussed here work well when the noise is stationary and where an optimal pre-whitener can be found from measurements when only the noise is present, but as already mentioned this may not be the case for speech and audio signals. It may still be possible, though, to apply a fixed pre-whitening filter that works well on average. The optimal pre-whitener, though, is obviously adaptive. An example of a pre-whitening system for audio signals can be found in [137]. It should also be noted that, although described here in the context of subspace methods, the principle can be applied to also the statistical methods and the filtering methods to obtain simpler estimators.

4.5   RANK ESTIMATION USING EIGENVALUES

We will now show how to determine the dimensions of the signal and noise subspaces from the eigenvalues of the sample covariance matrix, i.e., the ranks of the matrices $\mathbf{Z}$ and $\mathbf{G}$. In the case of single-pitch estimation, this amounts to determining the number of harmonics, since $\text{rank}(\mathbf{S}) = L_k$, while for the multi-pitch case, it is equivalent to estimating the total number of harmonics for all sources as $\text{rank}(\mathbf{S}) = \sum_{k=1}^{K} L_k = L$. We will do this by first deriving the log-likelihood function based on the eigenvalues of the covariance matrix and later show how to use this for finding the subspace dimension. Assuming that our signal vector $\mathbf{x}(n)$ is Gaussian and zero-mean, its likelihood function is given by

$$p(\mathbf{x}(n); \boldsymbol{\zeta}) = \frac{1}{\pi^M \det(\mathbf{R})} e^{-\mathbf{x}^H(n)\mathbf{R}^{-1}\mathbf{x}(n)}, \tag{4.33}$$

where $\boldsymbol{\zeta}$ contains our parameter set, here consisting of the signal subspace eigenvalues and eigenvectors along with the noise variance. For a collection of G such independent vectors, the likelihood function is

$$p(\{\mathbf{x}(n)\}; \boldsymbol{\zeta}) = \prod_{n=0}^{G-1} p(\mathbf{x}(n); \boldsymbol{\zeta}) \tag{4.34}$$

$$= \frac{1}{\pi^{GM} \det(\mathbf{R})^G} e^{-\sum_{n=0}^{G} \mathbf{x}^H(n)\mathbf{R}^{-1}\mathbf{x}(n)}. \tag{4.35}$$

By taking the logarithm, we obtain the log-likelihood function

$$\mathcal{L}(\boldsymbol{\zeta}) = \ln p(\{\mathbf{x}(n)\}; \boldsymbol{\zeta}) \tag{4.36}$$

$$= -GM \ln \pi - G \ln \det(\mathbf{R}) - \sum_{n=0}^{G-1} \mathbf{x}^H(n)\mathbf{R}^{-1}\mathbf{x}(n). \tag{4.37}$$

By noting that $\mathbf{x}^H(n)\mathbf{R}^{-1}\mathbf{x}(n) = \text{Tr}\{\mathbf{x}^H(n)\mathbf{R}^{-1}\mathbf{x}(n)\}$, we may rewrite this into

$$\mathcal{L}(\boldsymbol{\zeta}) = -GM \ln \pi - G \ln \det(\mathbf{R}) - \text{Tr}\left\{\sum_{n=0}^{G-1} \mathbf{x}^H(n)\mathbf{R}^{-1}\mathbf{x}(n)\right\} \tag{4.38}$$

$$= -GM \ln \pi - G \ln \det(\mathbf{R}) - \text{Tr}\left\{\mathbf{R}^{-1} \sum_{n=0}^{G-1} \mathbf{x}(n)\mathbf{x}^H(n)\right\}. \tag{4.39}$$

Next, we identify $\sum_{n=0}^{G-1} \mathbf{x}(n)\mathbf{x}^H(n)$ as the sample covariance matrix $\widehat{\mathbf{R}}$ scaled by G, whereby we get

$$\mathcal{L}(\boldsymbol{\zeta}) = -GM \ln \pi - G \ln \det(\mathbf{R}) - G \, \text{Tr}\left\{\mathbf{R}^{-1}\widehat{\mathbf{R}}\right\}. \tag{4.40}$$

From the derivations of the covariance matrix model and its eigenvalue decomposition in the previous section, we know that for a candidate signal subspace dimension L', the covariance matrix can be written as

$$\mathbf{R}' = \sum_{v=1}^{L'} (\lambda_v - \sigma^2) \mathbf{u}_v \mathbf{u}_v^H + \sigma^2 \mathbf{I}. \tag{4.41}$$

The maximum likelihood estimates of the eigenvalues and the eigenvectors of (4.41) are obtained from the EVD of the sample covariance matrix [5] (see also [13]), here denoted $\{\hat{\lambda}_v\}$ and $\{\hat{\mathbf{u}}_v\}$, respectively. For a candidate signal subspace dimension L', we readily obtain an estimate of the noise variance from the eigenvalues as

$$\hat{\sigma}^2 = \frac{1}{M - L'} \sum_{v=L'+1}^{M} \hat{\lambda}_v. \tag{4.42}$$

Substituting these into our expression for the log-likelihood function, we get

$$\mathcal{L}(\zeta) = -GM \ln \pi - G \ln \det(\mathbf{R}') - G \operatorname{Tr}\left\{ \mathbf{R}'^{-1} \widehat{\mathbf{R}} \right\} \tag{4.43}$$
$$= -GM \ln \pi - G \ln \det(\mathbf{R}') - GM, \tag{4.44}$$

since

$$\operatorname{Tr}\left\{ \mathbf{R}'^{-1} \widehat{\mathbf{R}} \right\} = L + \sum_{v=L'+1}^{M} \frac{\hat{\lambda}_v}{\hat{\sigma}^2} = L + \frac{\sum_{v=L'+1}^{M} \hat{\lambda}_v}{\frac{1}{M-L'} \sum_{v=L'+1}^{M} \hat{\lambda}_v} \tag{4.45}$$
$$= L + (M - L) \frac{\sum_{v=L'+1}^{M} \hat{\lambda}_v}{\sum_{v=L'+1}^{M} \hat{\lambda}_v} = M \tag{4.46}$$

and finally

$$\mathcal{L}(\zeta) = -GM \ln \pi - G \ln \left(\prod_{v=1}^{L'} \hat{\lambda}_v \prod_{v=L'+1}^{M} \hat{\sigma}^2 \right) - GM \tag{4.47}$$
$$= -GM \ln \pi - G \ln \left(\prod_{v=1}^{L'} \hat{\lambda}_v \left(\frac{1}{M - L'} \sum_{v=L'+1}^{M} \hat{\lambda}_v \right)^{M-L'} \right) - GM, \tag{4.48}$$

which is the log-likelihood function expressed as in [13]. To arrive at the expression used in [177], we first observe that the determinant may also be written as

$$\prod_{v=1}^{L'} \hat{\lambda}_v = \frac{\prod_{v=1}^{M} \hat{\lambda}_v}{\prod_{v=L'+1}^{M} \hat{\lambda}_v} = \frac{\det(\mathbf{R}')}{\prod_{v=L'+1}^{M} \hat{\lambda}_v}. \tag{4.49}$$

Inserting this into the log-likelihood function leads us to the following:

$$\mathcal{L}(\boldsymbol{\zeta}) = -GM \ln \pi - G \ln \left(\frac{\prod_{v=1}^{M} \hat{\lambda}_v}{\prod_{v=L'+1}^{M} \hat{\lambda}_v} \left(\frac{1}{M-L'} \sum_{v=L'+1}^{M} \hat{\lambda}_v \right)^{M-L'} \right) - GM \qquad (4.50)$$

$$= -GM \ln \pi - G \ln \prod_{v=1}^{M} \hat{\lambda}_v - G(M-L') \ln \frac{\frac{1}{M-L'} \sum_{v=L'+1}^{M} \hat{\lambda}_v}{\prod_{v=L'+1}^{M} \hat{\lambda}_v^{1/(M-L')}} - GM. \qquad (4.51)$$

Since $\prod_{v=1}^{M} \hat{\lambda}_v$ is constant the log likelihood function depends only on the ratio between the arithmetic and geometric means of the eigenvalues $\{\hat{\lambda}_v\}_{v=L'+1}^{M}$ as shown in [177]. This expression can now be used for determining the dimension of the signal subspace by combining the log-likelihood function with an appropriate penalty function such as the AIC or the MDL criterion. But before we can do this, we need to determine the number of free parameters in our estimation problem (see [177] for details on this). Here, these consist of the L' real eigenvalues, the noise variance, and L' complex eigenvectors of length M. This amounts to $L' + 1 + 2ML'$ degrees of freedom, but due to the eigenvectors being constrained to unit norm and orthogonal, we are left with $L'(2M - L') + 1$ degrees of freedom. Using the AIC or MDL in combination with (4.51), we obtain the following cost function for determining the dimension of the signal subspace:

$$J(L') = -\mathcal{L}(\boldsymbol{\zeta}) + (L'(2M - L') + 1)\nu \qquad (4.52)$$

where

$$\nu = \begin{cases} 2 & \text{for} \quad \text{AIC} \\ \frac{1}{2} \ln N & \text{for} \quad \text{MDL} \end{cases} \qquad (4.53)$$

By minimizing this cost function, the signal subspace dimension and thus the number of sinusoids can be determined from the eigenvalues of the sample covariance matrix. This is illustrated in Figures 4.1 and 4.2 where the eigenvalues of a typical sample covariance matrix are shown for a signal containing five harmonics and the associated log-ratio between the arithmetic and geometric means of the eigenvalues and the MDL cost function are shown, respectively. Note that additive constants have been ignored.

It should be noted that AIC results in overestimation of the model order, meaning that it will result in a model order that is higher than the true one. On the other hand, it can be shown that the MDL is consistent [108], but both methods may fail in the presence of colored noise [178]. We refer the interested reader to [108] for a detailed discussion and analysis of the behavior of such criteria.

4.6   ANGLES BETWEEN SUBSPACES

As we have seen, the orthogonality property states that for the true parameters, the matrix $\mathbf{Z} \in \mathbb{C}^{M \times L}$ is orthogonal to the noise subspace eigenvectors in $\mathbf{G} \in \mathbb{C}^{M \times M-L}$. The question is then

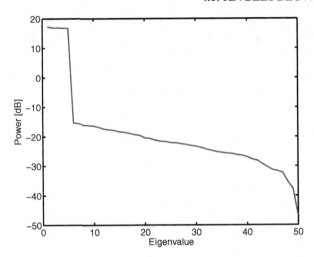

Figure 4.1: Empirical eigenvalues of a typical sample covariance matrix. The signal subspace has dimension five in this example.

how to measure the extend to which this relation holds. The concept of orthogonality is, of course, intimately related to the concept of angles, but it is a non-trivial question how to define angles in multiple dimensions. We will now proceed to derive an appropriate measure based on the angles between subspaces as known from linear algebra. The principal angles $\{\theta_k\}$ between the two subspaces $\mathcal{Z} = \mathcal{R}(\mathbf{Z})$ and $\mathcal{G} = \mathcal{R}(\mathbf{G})$ are defined recursively for $k = 1, \ldots, K$ as [64]

$$\cos(\theta_k) = \max_{\mathbf{u} \in \mathcal{Z}} \max_{\mathbf{v} \in \mathcal{G}} \frac{\mathbf{u}^H \mathbf{v}}{\|\mathbf{u}\|_2 \|\mathbf{v}\|_2} = \mathbf{u}_k^H \mathbf{v}_k, \tag{4.54}$$

where K is the minimal dimension of the two subspaces, i.e., $K = \min\{L, M - L\}$ and $\mathbf{u}^H \mathbf{u}_i = 0$ and $\mathbf{v}^H \mathbf{v}_i = 0$ for $i = 1, \ldots, k - 1$. Moreover, the normalization in the denominator of (4.54) is included to ensure that the vectors have unit norm.

The principal angles are usually defined using orthonormal bases for the two subspaces [64]. Here, we will instead use their projection matrices, since these are uniquely defined and will be useful later on. The projection matrix for a subspace $\mathcal{X}$ is defined as

$$\mathbf{\Pi}_X = \mathbf{X} \left(\mathbf{X}^H \mathbf{X}\right)^{-1} \mathbf{X}^H. \tag{4.55}$$

We note that such projections matrices are Hermitian, i.e., $\mathbf{\Pi}_X^H = \mathbf{\Pi}_X$ and have the properties $\mathbf{\Pi}_X^m = \mathbf{\Pi}_X$ for $m = 1, 2, \ldots$ and $\|\mathbf{\Pi}_X\|_F^2 = \dim(\mathcal{X})$ where $\dim(\cdot)$ is the dimension of the subspace.

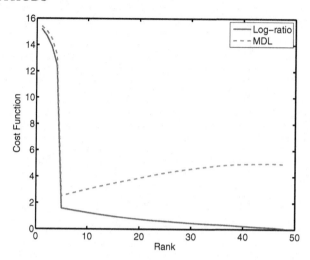

Figure 4.2: Log-ratio between the geometric and arithmetic means and the MDL cost function for the eigenvalues in Figure 4.1.

For subspace $\mathcal{G}$, the projection matrix is

$$\boldsymbol{\Pi}_G = \mathbf{G}\left(\mathbf{G}^H\mathbf{G}\right)^{-1}\mathbf{G}^H \tag{4.56}$$

$$= \mathbf{G}\mathbf{G}^H, \tag{4.57}$$

due to the columns of $\mathbf{G}$ being orthonormal, while, for subspace $\mathcal{Z}$, the projection matrix is

$$\boldsymbol{\Pi}_Z = \mathbf{Z}\left(\mathbf{Z}^H\mathbf{Z}\right)^{-1}\mathbf{Z}^H. \tag{4.58}$$

We can now write (4.54) into something more useful using the two projection matrices (4.57) and (4.58), i.e.,

$$\cos(\theta_k) = \max_{\mathbf{y}}\max_{\mathbf{z}} \frac{\mathbf{y}^H\boldsymbol{\Pi}_Z\boldsymbol{\Pi}_G\mathbf{z}}{\|\mathbf{y}\|_2\|\mathbf{z}\|_2} \tag{4.59}$$

$$= \mathbf{y}_k^H\boldsymbol{\Pi}_Z\boldsymbol{\Pi}_G\mathbf{z}_k = \kappa_k, \tag{4.60}$$

for $k = 1, \ldots, K$. We also require that the vectors are orthogonal, i.e., $\mathbf{y}^H\mathbf{y}_i = 0$ and $\mathbf{z}^H\mathbf{z}_i = 0$ for $i = 1, \ldots, k - 1$. It follows that $\{\kappa_k\}$ are the singular values of the matrix product $\boldsymbol{\Pi}_Z\boldsymbol{\Pi}_G$ ordered by decreasing value, and the two sets of vectors $\{\mathbf{y}\}$ and $\{\mathbf{z}\}$ are the left and right singular vectors of the matrix product, respectively. The corresponding principal angles, therefore, are ordered and bounded as

$$0 \leq \theta_1 \leq \ldots \leq \theta_K \leq \frac{\pi}{2}. \tag{4.61}$$

The singular values are related to the Frobenius norm of the product $\mathbf{\Pi}_Z \mathbf{\Pi}_G$ in the following way:

$$\|\mathbf{\Pi}_Z \mathbf{\Pi}_G\|_F^2 = \text{Tr}\left\{\mathbf{\Pi}_Z \mathbf{\Pi}_G \mathbf{\Pi}_G^H \mathbf{\Pi}_Z^H\right\} \tag{4.62}$$

$$= \text{Tr}\{\mathbf{\Pi}_Z \mathbf{\Pi}_G\} = \sum_{k=1}^{K} \kappa_k^2. \tag{4.63}$$

Interestingly, this can be related to the Frobenius norm of the difference between the two projection matrices, i.e.,

$$\|\mathbf{\Pi}_Z - \mathbf{\Pi}_G\|_F^2 = \text{Tr}\{\mathbf{\Pi}_Z + \mathbf{\Pi}_G - 2\mathbf{\Pi}_Z \mathbf{\Pi}_G\} \tag{4.64}$$

$$= L + (M - L) - 2\,\text{Tr}\{\mathbf{\Pi}_Z \mathbf{\Pi}_G\} \tag{4.65}$$

$$= M - 2\|\mathbf{\Pi}_Z \mathbf{\Pi}_G\|_F^2. \tag{4.66}$$

As can be seen, the minimization of (4.63) is the same as maximization of the Frobenius norm of the difference between the projection matrices. The Frobenius norm of the product $\mathbf{\Pi}_Z \mathbf{\Pi}_G$ can also be rewritten as

$$\|\mathbf{\Pi}_Z \mathbf{\Pi}_G\|_F^2 = \text{Tr}\left\{\mathbf{\Pi}_Z \mathbf{\Pi}_G \mathbf{\Pi}_G^H \mathbf{\Pi}_Z^H\right\} = \text{Tr}\left\{\mathbf{\Pi}_Z \mathbf{\Pi}_G^H\right\} \tag{4.67}$$

$$= \text{Tr}\left\{\mathbf{Z}\left(\mathbf{Z}^H \mathbf{Z}\right)^{-1} \mathbf{Z}^H \mathbf{G} \mathbf{G}^H\right\}. \tag{4.68}$$

This expression can be seen to be somewhat complicated, since it involves matrix inversion. It can be simplified by noting, as we have done numerous times, that the columns of $\mathbf{Z}$ consist of complex sinusoids that are asymptotically orthogonal for any distinct set of frequencies, i.e.,

$$\lim_{M \to \infty} M \mathbf{\Pi}_Z = \lim_{M \to \infty} M \mathbf{Z}\left(\mathbf{Z}^H \mathbf{Z}\right)^{-1} \mathbf{Z}^H \tag{4.69}$$

$$= \mathbf{Z} \mathbf{Z}^H. \tag{4.70}$$

Using this as an approximation for finite M, we can simplify (4.68) into a familiar form:

$$\|\mathbf{\Pi}_Z \mathbf{\Pi}_G\|_F^2 = \text{Tr}\left\{\mathbf{Z}\left(\mathbf{Z}^H \mathbf{Z}\right)^{-1} \mathbf{Z}^H \mathbf{G} \mathbf{G}^H\right\}$$

$$\approx \frac{1}{M} \text{Tr}\left\{\mathbf{Z}^H \mathbf{G} \mathbf{G}^H \mathbf{Z}\right\} = \frac{1}{M}\|\mathbf{Z}^H \mathbf{G}\|_F^2. \tag{4.71}$$

Except for the scaling $1/M$, this is the reciprocal of the original MUSIC cost function as introduced in [150]. By combining (4.63) and (4.71), we get

$$\frac{1}{M}\|\mathbf{Z}^H \mathbf{G}\|_F^2 \approx \sum_{k=1}^{K} \kappa_k^2 = \sum_{k=1}^{K} \cos^2(\theta_k). \tag{4.72}$$

This means that the original MUSIC cost function can be understood in the context of angles between subspaces. Curiously, this interpretation holds only for signal models consisting of vectors that are orthogonal or asymptotically orthogonal.

We now arrive at a measure of the extent to which the orthogonality property holds by averaging over all the nontrivial principal angles $K = \min\{L, M - L\}$ between $\mathcal{A}$ and $\mathcal{G}$ as

$$\frac{1}{K} \sum_{k=1}^{K} \cos^2(\theta_k) = \frac{1}{K} \sum_{k=1}^{K} \kappa_k^2 = \frac{1}{K} \|\boldsymbol{\Pi}_Z \boldsymbol{\Pi}_G\|_F^2 \tag{4.73}$$

$$\approx \frac{1}{MK} \|\mathbf{Z}^H \mathbf{G}\|_F^2 \triangleq J. \tag{4.74}$$

This measure has the desirable property that it is bounded and only zero when all angles are $\frac{\pi}{2}$, i.e., when the subspaces $\mathcal{A}$ and $\mathcal{B}$ are orthogonal in all directions. Additionally, the intersection of the subspaces is the range of the set of principal vectors for which $\cos(\theta_k) = 1$. The measure can easily be seen to be bounded as

$$0 \leq \frac{1}{K} \sum_{k=1}^{K} \cos^2(\theta_k) \leq 1. \tag{4.75}$$

If the division by K is left out, the measure is, in principle, unbounded. Note that the bound is also asymptotically valid for the right-most expression in (4.74).

It is also possible to express the measure in (4.74) in terms of the signal subspace eigenvectors. More specifically, due to the columns of $\mathbf{S}$ and $\mathbf{G}$ spanning complementary subspaces, we have that

$$\mathbf{I} = \mathbf{SS}^H + \mathbf{GG}^H. \tag{4.76}$$

Using this, we can rewrite (4.74) as

$$J = \frac{1}{MK} \|\mathbf{Z}^H \mathbf{G}\|_F^2 \tag{4.77}$$

$$= \frac{1}{MK} \operatorname{Tr}\left\{\mathbf{Z}^H \mathbf{GG}^H \mathbf{Z}\right\} = \frac{1}{MK} \operatorname{Tr}\left\{\mathbf{Z}^H \left(\mathbf{I} - \mathbf{SS}^H\right) \mathbf{Z}\right\} \tag{4.78}$$

$$= \frac{1}{MK} \left(\operatorname{Tr}\left\{\mathbf{Z}^H \mathbf{Z}\right\} - \operatorname{Tr}\left\{\mathbf{Z}^H \mathbf{SS}^H \mathbf{Z}\right\}\right) \tag{4.79}$$

$$= \frac{1}{MK} \left(ML - \|\mathbf{Z}^H \mathbf{S}\|_F^2\right), \tag{4.80}$$

where the last line follows from $\|\mathbf{Z}\|_F^2$ being ML. The relation extends beyond this measure. In fact, we could have redefined the angles between the subspaces using (4.76) and arrived at similar results, only the angle between the subspaces should now be minimized instead of maximized. It is generally beneficial in terms of computational complexity to use (4.80) instead of (4.77) whenever the dimension of the signal subspace is smaller than the dimension of the noise subspace (and vice versa).

The measure in (4.74) can be brought into a familiar form, similar to the statistical order estimation criteria like the MAP criterion in Section 2.6 by taking the logarithm of (4.74), i.e.,

$$\ln J = \ln \|\mathbf{Z}^H \mathbf{G}\|_F^2 - \ln(MK), \tag{4.81}$$

which can be seen to consist of two familiar terms: a "goodness of fit" measure and an order-dependent penalty function.

4.7   ESTIMATION USING ORTHOGONALITY

As mentioned previously, the MUSIC algorithm is based on the observation that $\mathcal{R}(\mathbf{G})$ $\perp \mathcal{R}(\mathbf{Z})$, i.e., that the matrix $\mathbf{Z}$ is orthogonal to the noise subspace. For the single-pitch case, the covariance matrix model is

$$\mathbf{R}_k = \mathrm{E}\left\{\mathbf{x}_k(n)\mathbf{x}_k^H(n)\right\} \tag{4.82}$$

$$= \mathbf{Z}_k \mathbf{P}_k \mathbf{Z}_k^H + \sigma^2 \mathbf{I}. \tag{4.83}$$

As we have seen, the noise subspace eigenvectors in the matrix $\mathbf{G}$ are then orthogonal to the columns of the matrix $\mathbf{Z}_k$, i.e., the complex sinusoids. This can be exploited for estimating the parameters of $\mathbf{Z}_k$, in our case a single parameter, the fundamental frequency ω_k, by forming a matrix $\mathbf{Z}_k$ for different candidate frequencies and measuring the angles between the subspaces, i.e.,

$$\hat{\omega}_k = \arg\min_{\omega_k} \|\mathbf{Z}_k^H \mathbf{G}\|_F^2 \tag{4.84}$$

$$= \arg\min_{\omega_k} \sum_{l=1}^{L_k} \|\mathbf{z}^H(\omega_k l)\mathbf{G}\|_2^2. \tag{4.85}$$

Or, in words, we find an estimate of the fundamental frequency by maximizing the angles between the subspaces $\mathcal{R}(\mathbf{Z}_k)$ and $\mathcal{R}(\mathbf{G})$. Note that the normalization in (4.74) is excluded here, since it is a constant. In Figure 4.3, an example of the cost function in (4.85) is shown for the signals in Figures 1.1 and 1.4, respectively. As can be seen, the fundamental frequencies can easily be identified as the global minima of the respective cost functions.

As we have discussed, the number of harmonics in $\mathbf{Z}_k$ and thus also the rank of $\mathbf{G}$ is generally unknown and has to be estimated on a segment-to-segment basis. This can be done using the eigenvalues as discussed earlier, but it is also possible to do so from the eigenvectors using the concept of angles between subspaces. The key observation is that the subspace properties discussed in Section 4.3 only hold when the matrices $\mathbf{Z}_k$, $\mathbf{S}$ and $\mathbf{G}$ are partitioned into subspaces having the right rank, i.e., $\mathbf{Z}_k$ will only be orthogonal to $\mathbf{G}$ and span the same space as $\mathbf{S}$ when the true model order is used. This is essentially because the sinusoids are linearly independent but not orthogonal, i.e., a subset of eigenvectors does not span the same space as a subset of sinusoids. By changing the estimator in (4.85) to include also the model order as an optimization parameter, we can take this

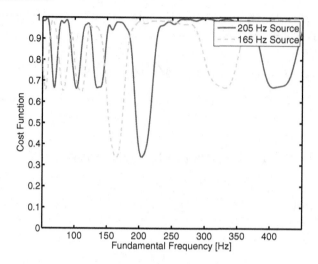

Figure 4.3: Cost function in (4.85) based on subspace orthogonality for the two signals in Figures 1.1 and 1.4, respectively.

into account, but now we have to include the normalization in (4.74), since this changes with the model order. Using (4.74), we thus arrive at the following expression

$$\hat{\omega}_k = \arg\min_{\omega_k} \min_{L_k} \frac{1}{MK} \| \mathbf{Z}_k^H \mathbf{G} \|_F^2 \tag{4.86}$$

$$= \arg\min_{\omega_k} \min_{L_k} \frac{1}{MK} \sum_{l=1}^{L_k} \| \mathbf{z}^H(\omega_k l) \mathbf{G} \|_2^2, \tag{4.87}$$

with $K = \min\{L_k, M - L_k\}$. Note that the model order is also found in the process. The interval over which to optimize L_k must be chosen with care and depends on the fundamental frequency, since, to avoid aliasing, $\omega_k L_k < 2\pi$. If the normalization is not used, the estimator will not work as the error will grow as a function of the matrix dimensions. Minimizing the cost function of (4.85) is equivalent to maximizing the average of the angles between the subspaces. Therefore, if the cost function is exactly zero, all angles between the two subspaces are $\frac{\pi}{2}$. An example of the cost function as a function of the model order is shown in Figure 4.4 for a signal containing five harmonics, i.e., $L_k = 5$. More specifically, the function shown is $\arg\min_{\omega_k} \frac{1}{MK} \sum_{l=1}^{L_k} \| \mathbf{z}^H(\omega_k l) \mathbf{G} \|_2^2$, which depends only on the model order. As can be seen, the true model order can be identified quite clearly from the curve. In Figure 4.5, estimates obtained using (4.87) are shown in the bottom panel for the clarinet signal whose spectrogram is depicted in the top panel. The estimates were obtained from 30 ms segments and with $M = N/2$. As can be seen, the estimates are rather accurate following the spectrogram as expected, even as the number of harmonics vary. It can also be seen that the signal is, in fact, a multi-pitch signal during the transitions from one note to another.

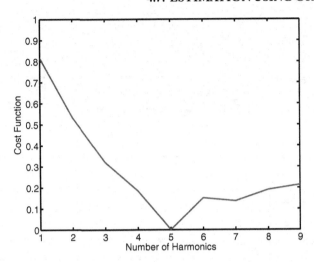

Figure 4.4: An example of the cost function based on subspace orthogonality as a function of the model order L_k for a signal where the true order is five.

We will now generalize the estimators to take into account the presence of multiple periodic components at the same time. To recapitulate, the covariance matrix model for the multi-pitch case is

$$\mathbf{R} = \sum_{k=1}^{K} \mathbf{Z}_k \mathbf{P}_k \mathbf{Z}_k^H + \sigma^2 \mathbf{I} \tag{4.88}$$

$$= \mathbf{Z}\mathbf{P}\mathbf{Z}^H + \sigma^2 \mathbf{I}. \tag{4.89}$$

The subspace orthogonality property states that the matrix $\mathbf{Z}$ and all its sub-matrices are orthogonal to $\mathbf{G}$, i.e.,

$$\mathbf{Z}^H \mathbf{G} = \mathbf{0} \quad \text{and} \quad \mathbf{Z}_k^H \mathbf{G} = \mathbf{0} \quad \forall k. \tag{4.90}$$

Note that the rank of $\mathbf{G}$ is related to the rank of $\mathbf{Z}$ and thus depends on all the sub-matrices $\mathbf{Z}_k$. First, assume that the model orders are known and thus also the dimensions of the subspaces. Writing out the formula for the angles between the subspaces spanned by the columns of $\mathbf{Z}$ and $\mathbf{G}$, we get

$$\|\mathbf{Z}^H \mathbf{G}\|_F^2 = \sum_{k=1}^{K} \|\mathbf{Z}_k^H \mathbf{G}\|_F^2, \tag{4.91}$$

which shows the cost function is additive over the sub-matrices associated with the individual sources. The estimates of the set of fundamental frequencies that lead to orthogonal matrices $\mathbf{Z}$ and $\mathbf{G}$ are

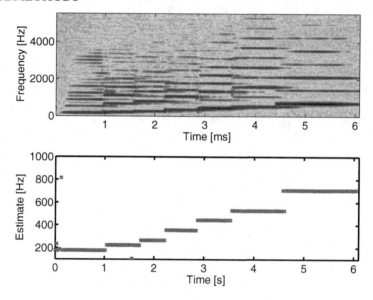

Figure 4.5: Clarinet signal spectrogram (top) and pitch estimates obtained using the joint estimator (bottom).

thus

$$\{\hat{\omega}_k\} = \arg\min_{\{\omega_k\}} \|\mathbf{Z}^H\mathbf{G}\|_F^2 = \arg\min_{\{\omega_k\}} \sum_{k=1}^{K} \|\mathbf{Z}_k^H\mathbf{G}\|_F^2 \qquad (4.92)$$

$$= \arg \sum_{k=1}^{K} \min_{\omega_k} \|\mathbf{Z}_k^H\mathbf{G}\|_F^2, \qquad (4.93)$$

where the last line follows from the minimization being independent for each source k. This facilitates estimation optimization of the cost function independently for the individual sources, i.e.,

$$\hat{\omega}_k = \arg\min_{\omega_k} \|\mathbf{Z}_k^H\mathbf{G}\|_F^2 \quad \forall k. \qquad (4.94)$$

It can be seen that the problem is effectively decoupled into a number of simpler problems. This is a notable feature of the subspace method. Unlike the optimal maximum likelihood estimator, it leads naturally to a set of optimization problems that each depend only on one single nonlinear parameter, ω_k. The fact, or assumption, that leads to this, is that $\mathrm{E}\left\{\mathbf{a}_k(n)\mathbf{a}_k(n)^H\right\} = \mathbf{P}_k$ is diagonal. In Figure 4.6, the cost function of the subspace-based multi-pitch estimator in (4.94) is depicted. As can be seen, it does lead to accurate estimates, but the cost function can also be seen to be much more complicated than for the single-pitch case.

For the case where the model orders $\{L_k\}$ and thus the rank of $\mathbf{G}$ are unknown, the problem becomes more complicated, but can still be solved within this framework. As before, we include an

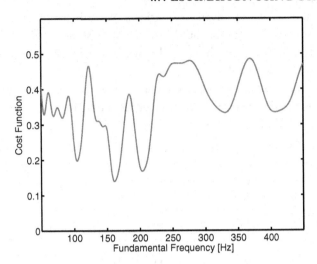

Figure 4.6: Cost function (4.94) based on subspace orthogonality for a mixture of the two signals in Figures 1.1 and 1.4 having fundamental frequencies 205 and 165 Hz, respectively.

optimization over the orders $\{L_k\}$ to obtain the set of fundamental frequencies as

$$\{\hat{\omega}_k\} = \arg\min_{\{\omega_k\}} \min_{\{L_k\}} \frac{1}{MK} \|\mathbf{Z}^H \mathbf{G}\|_F^2 \tag{4.95}$$

$$= \arg\min_{\{\omega_k\}} \min_{\{L_k\}} \sum_{k=1}^{K} \frac{1}{MK} \|\mathbf{Z}_k^H \mathbf{G}\|_F^2, \tag{4.96}$$

where $K = \min\{\sum_{k=1}^{K} L_k, M - \sum_{k=1}^{K} L_k\}$, since the dimensions of the signal and noise subspaces depend on the total number of harmonics. We see that the problem now requires an optimization over the set of fundamental frequencies $\{\omega_k\}$ and the orders $\{L_k\}$, and needless to say, this is a complicated problem. Fortunately, it is possible to simplify it somewhat, by realizing that given the set of orders $\{L_k\}$, the noise subspace eigenvectors in $\mathbf{G}$ are given too and the additivity of the cost function over the sub-matrices can be used, i.e.,

$$\min_{\{\omega_k\}} \min_{\{L_k\}} \frac{1}{MK} \sum_{k=1}^{K} \|\mathbf{Z}_k^H \mathbf{G}\|_F^2 = \min_{\{L_k\}} \sum_{k=1}^{K} \min_{\omega_k} \frac{1}{MK} \|\mathbf{Z}_k^H \mathbf{G}\|_F^2. \tag{4.97}$$

which facilitates optimization for each individual source given the orders.

An alternative to the joint optimization over the orders and fundamental frequency is to first determine the subspace dimensions using the unconstrained model whereby the total number of sinusoids can be found from the angles between the subspaces. However, the joint optimization procedure is generally preferable, and especially so for the single-pitch case, because it has been observed to be more robust to adverse conditions. Nevertheless, it may still lead to a significant simplification of the problem for the multi-pitch case, since the matrix $\mathbf{G}$ then is known (and fixed) throughout the optimization. Additionally, knowledge of the subspace dimensions also limit the optimization range for the set of orders $\{L_k\}$.

The major sources of computational complexity of the subspace method are the computation of the EVD of the covariance matrix, the matrix product $\mathbf{Z}_k^H \mathbf{G}$, and the evaluation of the Frobenius norm. We will now show how the last two steps can be implemented efficiently using FFTs. First, we define the Fourier matrix $\mathbf{F} \in \mathbb{C}^{F \times F}$, with $F \gg N$, as

$$
\mathbf{F} = \begin{bmatrix} 1 & 1 & 1 & \cdots & 1 \\ 1 & z^1 & z^2 & \cdots & z^{F-1} \\ \vdots & \vdots & \vdots & & \vdots \\ 1 & z^{(F-1)} & z^{2(F-1)} & \cdots & z^{(F-1)(F-1)} \end{bmatrix},
\tag{4.98}
$$

where $z = e^{j2\pi \frac{1}{F}}$. Next, we define a matrix $\mathbf{D} \in \mathbb{R}^{F \times M}$ containing the squared absolute values of the inverse FFTs of the zero-padded eigenvectors in $\mathbf{U}$ as

$$
[\mathbf{D}]_{lm} = \left| \left[\mathbf{F}^H \begin{bmatrix} \mathbf{U} \\ \mathbf{0} \end{bmatrix} \right]_{lm} \right|^2 ,
\tag{4.99}
$$

with $[\mathbf{D}]_{lm}$ being the (l, m)th element of $\mathbf{D}$. For a candidate pair of a fundamental frequency of $2\pi \frac{f}{F}$ and an order L_k, the Frobenius norm can be calculated as

$$
\|\mathbf{Z}_k^H \mathbf{G}\|_F^2 = \sum_{m=L_k+1}^{M} \sum_{l=1}^{L_k} [\mathbf{D}]_{(fl+1)m} .
\tag{4.100}
$$

Thus, the complexity of calculating $\|\mathbf{Z}^H \mathbf{G}\|_F^2$ for different ω_k and L_k can be significantly reduced by calculating the inverse FFT of all the eigenvectors once for each given data set. We note that some of the eigenvectors, corresponding to the largest eigenvalues, can be excluded from definition of (4.99), since there is a lower bound on L_k. Here, we have included all of them for notational simplicity. The FFT approach can then be combined with numerical optimization methods in order to obtain estimates that are not restricted to the FFT bins. For more on this, we refer the interested reader to [21, 23, 33].

Regarding the calculation of the EVD of the covariance matrix, it is possible to update it recursively over time using so-called subspace trackers (see, e.g., [36, 165]), which leads to a considerable reduction of the required computation time. The computation of the EVD has complexity $\mathcal{O}(M^3)$, while the most efficient subspace trackers are able to reduce the complexity to $\mathcal{O}(ML)$, with L being the dimension of the signal subspace to be tracked, i.e., $L = \sum_{k=1}^{K} L_k$, in the most extreme case. There is, however, one caveat: if the order is unknown and is determined using the angles between subspaces, then the subspace tracker must track the eigenvectors and not just an arbitrary basis of the space. The reason is this: a sub-partitioning of the basis is not generally equivalent to the sub-partitioning of the eigenvectors (see [85] for a detailed discussion of this). Some examples of subspace trackers that track only an arbitrary basis are PAST [179], NIC [118], FAPI [8], and FDPM [45], while the PASTd [179], SWASVD3 [10], LORAF3 [165], and FST [138] are examples of subspace trackers that track the eigenvectors and are thus applicable to the problem of determining the model order from the eigenvectors. It should be noted that for the multi-pitch case, the entire signal subspace has to be tracked.

4.8   ROBUST ESTIMATION

We will now show how to take into account that the harmonic frequencies are not exact integer multiples of a fundamental frequency. We will do this for the single-pitch case, i.e., we will operate on the signal $x_k(n)$ and also assume that the number of harmonics L_k is known or found a priori. It is, however, straightforward to relax these assumptions. We are here concerned with a set of sinusoids having frequencies $\{\psi_{k,l}\}$ corrupted by an additive white complex circularly symmetric Gaussian noise, $e_k(n)$, for $n = 0, \dots, N - 1$,

$$x_k(n) = \sum_{l=1}^{L} a_{k,l} e^{j\psi_{k,l}n} + e_k(n), \tag{4.101}$$

i.e., we will use the unconstrained model. For the perfectly harmonic case, the frequencies of the harmonics are exact integer multiples of a fundamental frequency ω_k, i.e., $\psi_{k,l} = \omega_k l$ with $l \in \mathbb{N}$, in which case the signal model in (4.101) is characterized by a single nonlinear parameter. As already discussed, this model is not always a good fit. Depending on the instrument, different parametric models of the inharmonicity of the harmonics can be derived from physical models (see, e.g., [54]). An example of such a model for stiff-stringed instruments is $\psi_{k,l} = \omega_k l \sqrt{1 + B_k l^2}$ where $B_k \ll 1$ is an unknown, positive stiffness parameter. Allowing for this deviation, the model in (4.101) now contains two unknown nonlinear parameters, namely ω_k and B_k, which complicates matters somewhat. Instead, we will here use a model where the frequencies $\{\omega_l\}$ are modeled as $\omega_l = \omega_0 l + \Delta_l$ with $\{\Delta_l\}$ being a set of small unknown perturbations that are to be estimated along with the fundamental frequency ω_0. We refer to this frequency model as the perturbed model. The perturbations $\{\Delta_l\}$ should be small, since arbitrarily large perturbations will result in meaningless estimates of ω_0.

Based on the angles between the subspaces, the fundamental frequency ω_k and the stiffness parameter B_k of the parametric inharmonicity model can be estimated in a straight-forward manner. Specifically, the $\mathbf{Z}_k$ matrix is constructed from the two parameters as

$$\mathbf{Z}_k = \begin{bmatrix} \mathbf{z}(\omega_k\sqrt{1 + B_k}) & \cdots & \mathbf{z}(\omega_k L_k\sqrt{1 + B_k L_k^2}) \end{bmatrix}. \tag{4.102}$$

Using this matrix, we can obtain estimates of the fundamental frequency and stiffness parameter as

$$(\hat{\omega}_k, \hat{B}_k) = \arg\min_{\omega_k, B_k} \|\mathbf{Z}_k^H \mathbf{G}\|_F^2, \tag{4.103}$$

which has to be evaluated for a large range of combinations of the two parameters. Here, the matrix $\mathbf{G}$ can be the noise subspace of either the covariance matrix of the single-pitch or multi-pitch signals, whichever is available.

The more interesting question is how the fundamental frequency and the individual frequencies can be found for the perturbed model where $\psi_{k,l} = \omega_k l + \Delta_{k,l}$, i.e., the Vandermonde matrix containing the complex sinusoids is now characterized by ω_k and $\{\Delta_{k,l}\}$ as

$$\mathbf{Z}_k = \left[\begin{array}{ccc} \mathbf{z}(\omega_k + \Delta_{k,l}) & \cdots & \mathbf{z}(\omega_k L_k + \Delta_{k,L}) \end{array} \right]. \tag{4.104}$$

Regrettably, direct minimization of the MUSIC cost function will not only be very computationally demanding, we now have $L_k + 1$ nonlinear parameters, but it will also not lead to any meaningful estimates of the fundamental frequency as we have no control over the distribution of the perturbations $\{\Delta_{k,l}\}$. Instead, the idea employed here is to estimate the parameters by redefining the cost function as

$$J(\omega_k, \{\Delta_{k,l}\}) = \text{Tr}\left\{\mathbf{Z}_k^H \mathbf{G}\mathbf{G}^H \mathbf{Z}_k\right\} + P(\{\Delta_{k,l}\}), \tag{4.105}$$

where $P(\cdot)$ is a penalty function; here a non-decreasing function of a metric with $P(\{0\}) = 0$. Also, it is desirable that the penalty function is additive over the harmonics. Therefore, a natural choice is $P(\{\Delta_{k,l}\}) = \sum_{l=1}^{L_k} v_{k,l} |\Delta_{k,l}|^p$ with $p \geq 1$ which penalizes large perturbations $\Delta_{k,l}$. Also, $\{v_{k,l}\}$ is a set of positive regularization constants. To simplify the estimator, we note that the Frobenius norm is additive over the columns of $\mathbf{Z}_k$, i.e.,

$$J(\omega_k, \{\Delta_{k,l}\}) = \text{Tr}\left\{\mathbf{Z}_k^H \mathbf{G}\mathbf{G}^H \mathbf{Z}_k\right\} + \sum_{l=1}^{L_k} v_{k,l} |\Delta_{k,l}|^p \tag{4.106}$$

$$= \sum_{l=1}^{L_k} \mathbf{z}^H(\omega_k l + \Delta_{k,l}) \mathbf{G}\mathbf{G}^H \mathbf{z}(\omega_k l + \Delta_{k,l}) + \sum_{l=1}^{L_k} v_{k,l} |\Delta_{k,l}|^p. \tag{4.107}$$

Furthermore, by substituting $\psi_{k,l}$ by $\omega_k l + \Delta_{k,l}$ and $\Delta_{k,l}$ by $\psi_{k,l} - \omega_k l$ in (4.107), we get the following cost function

$$J(\omega_k, \{\psi_{k,l}\}) = \sum_{l=1}^{L_k} \mathbf{z}^H(\psi_{k,l}) \mathbf{G}\mathbf{G}^H \mathbf{z}(\psi_{k,l}) + v_{k,l} |\omega_{k,l} - \omega_k l|^p. \tag{4.108}$$

It can be seen that the first term no longer depends on the fundamental frequency or the perturbations but only on the frequency of the lth harmonic $\psi_{k,l}$. Moreover, the first term in (4.108) can be seen to be the reciprocal of the MUSIC pseudo-spectrum, which, conveniently, has to be calculated only once for each segment. It can also be seen that the cost function is additive over independent terms and, therefore, the minimization of the cost function can be performed independently for each harmonic. The fundamental frequency estimate can be rewritten as follows:

$$\hat{\omega}_k = \arg\min_{\omega_k} \min_{\{\psi_{k,l}\}} \left\{ \sum_{l=1}^{L_k} \mathbf{z}^H(\psi_{k,l}) \mathbf{G}\mathbf{G}^H \mathbf{z}(\psi_{k,l}) + v_{k,l} |\psi_{k,l} - \omega_k l|^p \right\}$$

$$= \arg\min_{\omega_k} \sum_{l=1}^{L_k} \min_{\psi_{k,l}} \left\{ \mathbf{z}^H(\psi_{k,l}) \mathbf{G}\mathbf{G}^H \mathbf{z}(\psi_{k,l}) + v_{k,l} |\psi_{k,l} - \omega_k l|^p \right\},$$

where the frequencies $\{\psi_{k,l}\}$ and thus perturbations are also found implicitly. It should now also be clear why it is beneficial that the penalty function is additive over the harmonics. This estimator is expected to be more robust towards model mismatch than the ideal model and the parametric inharmonicity model. For a given $\hat{\omega}_k$, the frequencies can simply be found for $l = 1, \ldots, L_k$ as

$$\hat{\omega}_k = \arg\min_{\psi_{k,l}} \left\{ \mathbf{z}^H(\psi_{k,l}) \mathbf{G}\mathbf{G}^H \mathbf{z}(\psi_{k,l}) + v_{k,l} |\psi_{k,l} - \hat{\omega}_k l|^p \right\}, \tag{4.109}$$

where only the penalty term changes over l. The regularization constants can be interpreted in the following ways:

- For large regularization constants $v_{k,l}$, the ensuing perturbation will be small and the model reduces to the perfectly harmonic case, whereas for $v_{k,l}$ close to zero, the estimates reduces to unconstrained frequencies, from which no meaningful fundamental frequency estimate can be found.

- The combination of a log-likelihood function with a penalty term is equivalent to a MAP approach with the penalty term being a log-prior on the perturbations. For a Gaussian prior, for example, we would have $p = 2$ and $v_{k,l} = 1/(2\rho_{k,l}^2)$ with $\rho_{k,l}^2$ being the variance of the perturbations of the lth harmonic (see Section 2.5 for more on MAP estimation).

- The regularization constants $v_{k,l}$ can also be interpreted as Lagrange multipliers. This means that the estimator can be thought of as a constrained estimator with a set of implicit constraints. In this sense, the method is conceptually related to the robust Capon beamformer of [164], which is based on explicit constraints.

It may be worth modifying the penalty function such that less emphasis is put on perturbations for higher harmonics or even use an asymmetrical penalty function, since the parametric inharmonicity model suggest that the harmonics will be higher than the integer multiple of the fundamental. This can be realized by, for example, using a piecewise linear norm (see, e.g., [14]). The main difficulty of the approach discussed in this section is that one needs to pick the regularization constants, something that may be difficult to do in practice.

4.9   ESTIMATION USING SHIFT-INVARIANCE

As we have seen, it is possible to estimate the dimension of the signal subspace using the eigenvalues of the sample covariance matrix. Interestingly, it is also possible to determine the rank of the matrices from the eigenvectors by exploiting certain properties of the signal subspace, namely the shift-invariance property. We will now review some methods for order estimation based on these principles, namely the ESTER [9] and SAMOS [127] methods. The ESTER and SAMOS methods are derived from the ESPRIT method [147], which is based on $\mathcal{R}\left(\mathbf{S}\right) = \mathcal{R}\left(\mathbf{Z}\right)$ and the shift-invariance property of the matrix $\mathbf{Z}$. The sinusoidal parameters are found using (4.21) by constructing the matrices $\underline{\mathbf{S}}$ and $\overline{\mathbf{S}}$ as shown in (4.20) and then solving for Ξ in

$$\overline{\mathbf{S}} \approx \underline{\mathbf{S}}\Xi, \tag{4.110}$$

in some sense. For instance,

$$\widehat{\Xi} = \arg\min_{\Xi} \|\overline{\mathbf{S}} - \underline{\mathbf{S}}\Xi\|_F^2 \tag{4.111}$$

$$= \left(\underline{\mathbf{S}}^H\underline{\mathbf{S}}\right)^{-1}\underline{\mathbf{S}}^H\overline{\mathbf{S}}, \tag{4.112}$$

where the sinusoidal frequencies are found from the eigenvalues of $\widehat{\Xi}$ via the relation in (4.22). The key observation is that equation (4.110) holds only when the eigenvectors of $\mathbf{R}$ are partitioned into a signal and a noise subspace such that the dimension of the signal subspace equals the true number of sinusoids. The argument is the same as before: the individual sinusoids in the columns of $\mathbf{Z}$ do not span the same space as individual eigenvectors and the same goes for subsets of sinusoids.

An order estimate can be obtained using the ESTER method in the following way: First, the sample covariance matrix and its EVD are found. Afterwards, the matrices $\underline{\mathbf{S}}$ and $\overline{\mathbf{S}}$ are constructed for each L from the EVD and the matrix Ξ is estimated. Finally, the order is found by measuring the goodness of the fit in (4.110) as

$$J(L) = \|\overline{\mathbf{S}} - \underline{\mathbf{S}}\widehat{\Xi}\|_2^2, \tag{4.113}$$

for various candidate orders L and then picking the one for which the modeling error is minimized in the sense of (4.113).

The SAMOS method works in a similar way. It is based on the rationale that the extent to which the relation in (4.110) holds can be measured from the singular values $\{\eta_k\}_{k=1}^{2L}$ of the augmented matrix $\Phi = \begin{bmatrix} \underline{\mathbf{S}} & \overline{\mathbf{S}} \end{bmatrix}$ for various L as

$$J(L) = \frac{1}{L}\sum_{k=L+1}^{2L} \eta_k^2, \tag{4.114}$$

since Φ will have rank L when the columns of $\underline{\mathbf{S}}$ can be described accurately as linear combinations of the columns in $\overline{\mathbf{S}}$. The order estimate is then found as the minimizer of the cost function (4.114).

This method requires that a singular value decomposition is calculated for each candidate order L. The SAMOS and ESTER methods are based on the same principle and essentially use the same measure and can, therefore, be expected to perform similarly. The ESTER method is, however, preferably in that it allows for a wider range of candidate orders and an efficient implementation exists as was shown in [9].

It is also possible to use the shift-invariance property of the signal model to estimate the fundamental frequency. The method presented next applies to the single-pitch case and $\mathbf{S}$, therefore, in the following, denotes the eigenvectors of the signal subspace of the covariance matrix for the kth source, i.e., $\mathbf{R}_k$ In practice, the expectation operator in (4.6) is replaced by a finite sum and the right relation in (4.21) holds only approximately and the underlying assumptions may only be approximations of the observed phenomenon. The sinusoidal parameters are, therefore, found by constructing the matrices $\underline{\mathbf{S}}$ and $\overline{\mathbf{S}}$ and then solving for $\boldsymbol{\Xi}$ in $\overline{\mathbf{S}} \approx \underline{\mathbf{S}}\boldsymbol{\Xi}$ like in (4.112) or using total least-squares. The sinusoidal frequencies are then obtained using the empirical EVD of $\widehat{\boldsymbol{\Xi}}$ via the relation in (4.22), i.e.,

$$\widehat{\boldsymbol{\Xi}} = \mathbf{C}\widehat{\mathbf{D}}\mathbf{C}^{-1} \tag{4.115}$$

with $\mathbf{C}$ containing the empirical eigenvectors of $\widehat{\boldsymbol{\Xi}}$ and

$$\widehat{\mathbf{D}} = \mathrm{diag}\left([\, e^{j\hat{\psi}_1} \, \cdots \, e^{j\hat{\psi}_{L_k}} \,] \right), \tag{4.116}$$

where $\{\hat{\psi}_v\}_{v=1}^{L_k}$ is a set of unconstrained frequencies. It is not clear how to estimate the fundamental frequency from these equations, since the eigenvalues are not constrained to be equally spaced on the unit circle. We proceed as follows: We order the eigenvalues and eigenvectors in (4.115) by increasing arguments, i.e., $\hat{\psi}_1 \leq \ldots \leq \hat{\psi}_{L_k}$. Using the shift-invariance property in (4.21) and (4.22), we can write $\overline{\mathbf{S}} = \underline{\mathbf{S}}\mathbf{B}^{-1}\mathbf{D}\mathbf{B}$, and thus

$$\overline{\mathbf{S}} \approx \underline{\mathbf{S}}\mathbf{C}\widehat{\mathbf{D}}\mathbf{C}^{-1}. \tag{4.117}$$

Defining the diagonal matrix containing the unknown fundamental frequency as

$$\widetilde{\mathbf{D}} = \mathrm{diag}\left([\, e^{j\omega_k} \, \cdots \, e^{j\omega_k L_k} \,] \right), \tag{4.118}$$

we introduce the cost function

$$J \triangleq \|\overline{\mathbf{S}} - \underline{\mathbf{S}}\mathbf{C}\widetilde{\mathbf{D}}\mathbf{C}^{-1}\|_F^2, \tag{4.119}$$

from which the fundamental frequency can be estimated as

$$\hat{\omega}_k = \arg\min_{\omega_k} J, \tag{4.120}$$

where only $\widetilde{\mathbf{D}}$ depends on ω_k. Note that also the order L_k can be estimated in this manner. It can be seen from (4.21) that in the ideal case, we have equality in (4.117). So, instead we may introduce the modified cost function as

$$J \triangleq \|\overline{\mathbf{S}}\mathbf{C} - \underline{\mathbf{S}}\mathbf{C}\widetilde{\mathbf{D}}\|_F^2 \tag{4.121}$$

$$= \|\mathbf{V} - \mathbf{W}\widetilde{\mathbf{D}}\|_F^2, \tag{4.122}$$

where $\mathbf{V} = \overline{\mathbf{S}}\mathbf{C}$ and $\mathbf{W} = \underline{\mathbf{S}}\mathbf{C}$. As $\mathbf{C}$ is not orthogonal, the minimization of this expression generally is not equivalent to minimizing (4.119). The cost function in (4.122) can be expanded as

$$J = -2\,\mathrm{Re}\left(\mathrm{Tr}\left\{\mathbf{V}\widetilde{\mathbf{D}}^H\mathbf{W}^H\right\}\right) \tag{4.123}$$

$$+ \mathrm{Tr}\left\{\mathbf{V}\mathbf{V}^H\right\} + \mathrm{Tr}\left\{\mathbf{W}\widetilde{\mathbf{D}}\widetilde{\mathbf{D}}^H\mathbf{W}^H\right\}. \tag{4.124}$$

It can be seen that the last two terms are constant and need not be included in the optimization, and we, therefore, introduce $\mathbf{H} = \mathbf{W}^H\mathbf{V}$ and redefine the cost function as

$$J \triangleq -2\,\mathrm{Re}\left(\mathrm{Tr}\left\{\mathbf{H}\widetilde{\mathbf{D}}^H\right\}\right) \tag{4.125}$$

$$= -2\,\mathrm{Re}\left(\sum_{l=1}^{L_k} h_l e^{-j\omega_k l}\right) \tag{4.126}$$

with $h_l = \left[\mathbf{H}\right]_{ll}$ and proceed as before and use (4.120) with this cost function for finding the fundamental frequency. The cost function in (4.126) can be seen to consist of a polynomial, i.e., $\sum_{l=1}^{L_k} h_l z^{-l}$. It is shown in Figure 4.7 as a function of the fundamental frequency for a signal having fundamental frequency 0.3142 and five harmonics. It can be seen that the cost function is very smooth, especially so compared to the cost function in Figure 4.3. Although it does not show here, the cost function is, actually, multi-modal and can contain multiple minima.

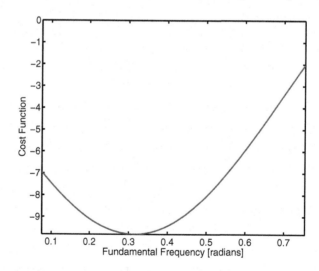

Figure 4.7: Cost function based on the shift-invariance of the signal subspace for a signal having fundamental frequency 0.3142.

We note that the method discussed here is superior to fitting the model matrix to the signal subspace using the relation in (4.18) and the MUSIC method in terms of computational complexity;

in fact, the method is quite simple and fast. The question is, however, how exactly the method differs from direct fitting of the fundamental frequency to the unconstrained frequencies in (4.117), i.e.,

$$\hat{\omega}_k = \arg \min_{\omega_k} \|\hat{\mathbf{D}} - \tilde{\mathbf{D}}\|_F^2, \tag{4.127}$$

To answer this question, we consider the ideal case, where the right relation in (4.21) holds exactly. In that case, we may write

$$\overline{\mathbf{S}} = \underline{\mathbf{S}}\hat{\Xi}. \tag{4.128}$$

Using this relation to rewrite (4.119), we get

$$J = \|\underline{\mathbf{S}}\hat{\Xi} - \underline{\mathbf{S}}\mathbf{C}\tilde{\mathbf{D}}\mathbf{C}^{-1}\|_F^2 \tag{4.129}$$
$$= \|\underline{\mathbf{S}}\mathbf{C}\hat{\mathbf{D}}\mathbf{C}^{-1} - \underline{\mathbf{S}}\mathbf{C}\tilde{\mathbf{D}}\mathbf{C}^{-1}\|_F^2 \tag{4.130}$$
$$= \|\underline{\mathbf{S}}\mathbf{C}\left(\hat{\mathbf{D}} - \tilde{\mathbf{D}}\right)\mathbf{C}^{-1}\|_F^2, \tag{4.131}$$

which is different from the original cost function as defined in (4.127). The WLS method of Section 2.10 is essentially a closed-form fit of the fundamental frequency to the unconstrained frequencies in a weighted least-squares sense, but it requires that the amplitudes are known or are estimated.

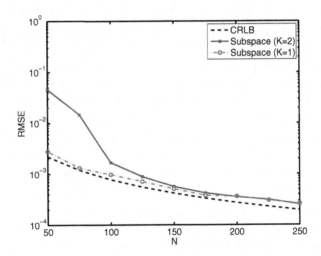

Figure 4.8: RMSE as a function of N with $PSNR = 10$ for one and two sources, respectively.

4.10 SOME RESULTS

We will now present some experimental results from Monte Carlo simulations based on synthetic signals. We will do this for the method based on subspace orthogonality and angles between subspaces

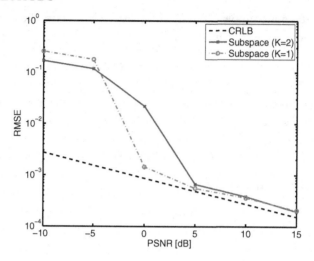

Figure 4.9: RMSE as a function of the PSNR with $N = 200$ for one and two sources, respectively.

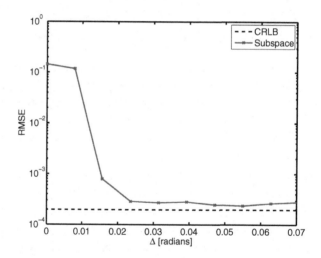

Figure 4.10: RMSE as a function of the difference between the fundamental frequencies of two source, i.e., $\Delta = |\omega_1 - \omega_2|$ for $N = 250$ and $PSNR = 10$.

for signals generated using the model in (1.5) in additive white Gaussian noise (for a discussion for the methods based on the shift-invariance property, see the next section). For each combination of parameters, 100 iterations were run and the RMSE was estimated from these. The experiments have been run for two sources, i.e., a multi-pitch signal, with the sources having fundamental frequencies 0.2964 and 0.2257, uniformly distributed phases, unit amplitudes and three harmonics, and for the single-pitch case consisting of only the source having fundamental frequency 0.2964. A sub-vector length of $N/2$ has been used throughout these experiments. In the first experiments, we assume that the model order is known and use the estimator in (4.94). In Figure 4.8, the RMSE is plotted as a function of the number of samples N, while the PSNR is held constant at 10 dB. Similarly, N is kept fixed at 200, while the PSNR is varied in the second experiment, with the results being shown in Figure 4.9. In both figures, the results are shown for both one and two sources. Note the CRLB shown here is the asymptotic CRLB, which is expected to hold for multiple sources as well when these are well-separated in frequency. For one source, i.e., for single-pitch signals, the method can be seen to perform well following the CRLB closely, although it does not attain it. When a second source is introduced, it can be seen that the estimator now exhibits thresholding behavior at low PSNRs and for low N, however this happens at very low N and PSNRs. In Figure 4.10, the RMSE is plotted as a function of the difference between the fundamental frequencies for the two sources, $\Delta = |\omega_1 - \omega_2|$, where ω_1 is kept fixed at 0.2964 while ω_2 is varied from 0.2257 to 0.2964 with $N = 250$ and $PSNR = 10$ dB. This experiment serves the purpose of testing the estimator under adverse conditions as closely spaced fundamental frequency will lead to interaction effects between the harmonics. Like the other algorithms (see Sections 2.11 and 3.10), the subspace-based method can be seen to perform quite well under these circumstances, below which the RMSE deviates from the CRLB by an order of magnitude, can be seen to be occur at quite low differences between the fundamental frequencies. Also, the ability of the method to estimate the correct model order has been investigated in simulations. The subspace-based method for estimating the model order is fundamentally different from those used in Sections 2.11 and 3.10 as these are both statistical methods. In this experiment, we run 1000 iterations and assume that the fundamental frequency is known, here 0.8170, and use the criterion in (4.87) for determining the model order for a single-pitch signal consisting of five unit-amplitude harmonics and uniformly distributed phases. The percentage of correctly estimated model order is shown as a function of N for $PSNR = 40$ dB in Figure 4.11 and as a function of the PSNR for $N = 500$ in Figure 4.12. The influence of the covariance matrix size M has also been investigated for $N = 200$ and $PSNR = 40$ dB with the results being shown in Figure 4.13. The method can be seen to perform extremely well, in fact, it performs just as well or better than the MAP-criterion (see Section 2.11) estimating the correct order with a very low probability of error. To summarize, the method based on the angles between subspaces performs well both with respect to fundamental frequency and order estimation.

4.11 DISCUSSION

As we have seen in this chapter, the problem of single- and multi-pitch estimation can be solved using subspace methods, and so can the associated problem of order estimation. In fact, several ways of doing this through the eigenvalues or eigenvectors of the covariance matrix exist. The shift-invariance principle of ESPRIT can be used for finding the fundamental frequency and the model order jointly. The resulting cost function turns out to be very smooth, compared to those of, e.g., NLS and MUSIC, as was observed in [22], and it is, therefore, both computationally and conceptually easy to perform numerical optimization using this principle. But, while ESPRIT is an extremely elegant

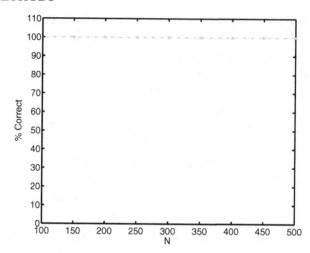

Figure 4.11: Percentage of correctly estimated model orders as a function of N with $PSNR = 40$ dB.

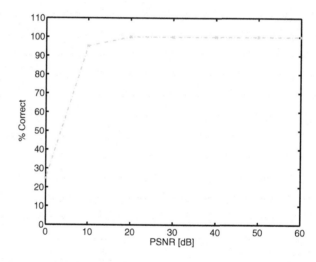

Figure 4.12: Percentage of correctly estimated model orders as a function of the PSNR with $N = 500$.

and conceptually simple solution to the problem of finding the frequencies of an unconstrained model, it is more complicated to apply it to pitch estimation and it does not easily lend itself to generalization to multi-pitch estimation. Also, it has been found to be very sensitive to spurious frequency estimates. Under good conditions, it does perform similarly to the MUSIC approach [22]. The MUSIC approach, which is based on the principle of measuring the angles between the noise subspace and the candidate model, on the other hand, has been demonstrated in [23] to be robust and provide high-resolution estimates of both the fundamental frequency and the model order for

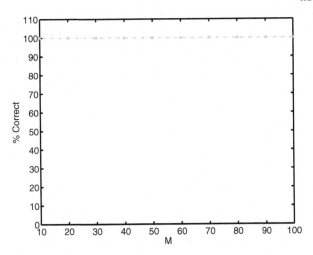

Figure 4.13: Percentage of correctly estimated model orders as a function of M for $N = 200$ and with $PSNR = 40$ dB.

the single-pitch case. Furthermore, for the multi-pitch case, the MUSIC approach leads to a natural decoupling of the multi-dimensional problem into a set of one-dimensional problems for the case of the model orders being known [33]. When the model orders are not known, the decoupling is only partial. A prominent feature of both the ESPRIT and MUSIC methods is that they offer a geometric alternative to finding the model order using statistical methods, something that, as we have seen, requires that many approximations be applied [44, 160]. It should be noted that the subspace methods are generally suboptimal as can be observed from the constant gap (in dB) between the RMSE curve and the CRLB [22, 23, 33], unlike the maximum likelihood methods of Chapter 2, but their performance increases as a function of the PSNR and the number of samples. In real applications, this is, however, less of an issue compared to that of robustness. The MUSIC approach has been seen to be less robust to adverse conditions, and it requires knowledge of the number of harmonics for all sources to work. For single-pitch estimation, the main drawback of the subspace methods is the computational complexity, but this can be lessened using subspace trackers and the use of FFTs and rootings methods. Also, the assumption that the observation noise is white may cause a problem, although we have observed this to be much less of an issue for the pitch estimation problem than the problem of finding unconstrained frequencies. Like for the filtering approaches, it is not straightforward to incorporate prior knowledge in the subspace methods, although certain relations exist between MUSIC and maximum likelihood estimation [157, 158].

CHAPTER 5

Amplitude Estimation

5.1 INTRODUCTION

Often, it is not only important to know the fundamental frequency of a periodic signal, but also to know the strength of such components. In some cases, even the phases of the individual harmonics may be of interest, for example, if the signal is to be reconstructed from the model or a modification thereof. The amplitude estimation problem is commonly recurring in a wide range of applications, and several parametric and non-parametric approaches to form such an estimate can be found in the literature. In this chapter, we will briefly review some such techniques, assuming that the frequencies of these periodic components have already been determined using one of the techniques discussed in the earlier chapters. As a result, we will here consider a slight generalization of the signal models given in (1.1) and (1.5), and instead study a signal consisting of L complex-valued sinusoids, namely the unconstrained model

$$x(n) = \sum_{l=1}^{L} a_l e^{j \psi_l n} + e(n), \tag{5.1}$$

for $n = 0, \ldots, N - 1$, where L as well as $\{\psi_l\}_{l=1}^{L}$ are assumed known, with $\psi_k \neq \psi_l$ for $k \neq l$, and where $e(n)$ denotes a zero mean, complex-valued, and assumed stationary (and possibly colored) additive noise. The complex amplitudes can also be expressed in terms of their (real) amplitude and phase as $a_l = A_l e^{j \phi_l}$. Thus, the problem of interest is given $\{x(n)\}_{n=0}^{N-1}$, estimate the complex amplitudes $\{a_l\}_{l=1}^{L}$. In the following, we begin by addressing this estimation problem assuming that the corrupting additive noise has a known covariance structure, thus allowing for the forming of a whitened signal. Then, we proceed to relax this often quite restrictive assumption, allowing for an unknown coloring of the additive noise, something that is particularly relevant for speech and audio signals.

5.2 LEAST-SQUARES ESTIMATION

We begin by examining the commonly used least-squares (LS) approaches to estimate the unknown amplitudes. Consider the observed, and possibly pre-whitened, signal $x(n)$, defined in (5.1). Then,

$$\begin{bmatrix} x(0) \\ \vdots \\ x(N-1) \end{bmatrix} = \begin{bmatrix} 1 & \cdots & 1 \\ e^{j \psi_1} & \cdots & e^{j \psi_L} \\ \vdots & \ddots & \vdots \\ e^{j \psi_1 (N-1)} & \cdots & e^{j \psi_L (N-1)} \end{bmatrix} \begin{bmatrix} a_1 \\ \vdots \\ a_L \end{bmatrix} + \begin{bmatrix} e(0) \\ \vdots \\ e(N-1) \end{bmatrix} \tag{5.2}$$

or, using a vector-matrix notation,

$$\mathbf{x} = \mathbf{Z}\mathbf{a} + \mathbf{e}. \tag{5.3}$$

The well-known LS estimate of the unknown amplitude vector is then

$$\hat{\mathbf{a}} = \left(\mathbf{Z}^H \mathbf{Z}\right)^{-1} \mathbf{Z}^H \mathbf{x}, \tag{5.4}$$

which is a statistically efficient estimator for all $N \geq L$ for white Gaussian noise [155]. As is well known, LS-approaches do not account for the (possibly existing) correlation of the additive noise, and are, therefore, in general, suboptimal. However, as shown in e.g. [155], if the noise is Gaussian, the above estimator is still asymptotically efficient, i.e., for sufficiently large data lengths, the variance of $\hat{\mathbf{a}}$ will reach the corresponding CRLB, given by (see, e.g., [152])

$$\mathrm{CRLB}(\hat{\mathbf{a}}) = \left(\mathbf{Z}^H \mathbf{Q}^{-1} \mathbf{Z}\right)^{-1}, \tag{5.5}$$

where $\mathbf{Q} = \mathrm{E}\{\mathbf{e}\mathbf{e}^H\}$, which for an additive unit variance white noise implies that $\mathbf{Q} = \mathbf{I}$. Alternatively, an approximate LS estimate may be formed from the L largest peaks of the DFT of $\{x(n)\}_{n=0}^{N-1}$, i.e.,

$$\hat{a}_l = \frac{1}{N} \sum_{n=0}^{N-1} x(n) e^{-j\psi_l n}, \quad \text{for} \quad l = 1, \ldots, L. \tag{5.6}$$

In [155], the authors show that this computationally much simpler approximate LS estimator is also statistically efficient asymptotically. The estimate in (5.4) often (but not always) yields preferable estimates, i.e., having lower mean-squared error (MSE), as compared to the estimate in (5.6).

5.3   CAPON- AND APES-LIKE AMPLITUDE ESTIMATES

In general, the additive noise can not be assumed to be white or to have a predetermined noise covariance matrix. We now proceed to examine possibilities to form Capon- and APES-like amplitude estimators, allowing for an arbitrary coloring of the corrupting noise. Such estimators may be formed in several ways, but all share the common characteristic that they form a bank of narrowband filters designed such that the total power at the filter output is minimized whereas the frequency or frequencies of interest are constrained appropriately. Splitting the measured data set into $N - M + 1$ sub-vectors of length M allows us to define

$$\mathbf{x}(n) \triangleq \begin{bmatrix} x(n) & \ldots & x(n + M - 1) \end{bmatrix}^T \tag{5.7}$$

$$= \begin{bmatrix} 1 & \cdots & 1 \\ e^{j\psi_1} & \cdots & e^{j\psi_L} \\ \vdots & \ddots & \vdots \\ e^{j\psi_1(M-1)} & \cdots & e^{j\psi_L(M-1)} \end{bmatrix} \begin{bmatrix} a_1 e^{j\psi_1 n} \\ \vdots \\ a_L e^{j\psi_L n} \end{bmatrix} + \begin{bmatrix} e(n) \\ \vdots \\ e(n + M - 1) \end{bmatrix} \tag{5.8}$$

$$\triangleq \mathbf{Z}(n)\mathbf{a} + \mathbf{e}(n), \tag{5.9}$$

where

$$\mathbf{Z}(n) = \mathbf{Z} \begin{bmatrix} e^{j\psi_1 n} & & \\ & \ddots & \\ & & e^{j\psi_L n} \end{bmatrix} \triangleq \mathbf{Z}\mathbf{D}_n.$$ (5.10)

By then introducing a set of narrowband filters, $\mathbf{h}_l$, designed such that a (generic) frequency of interest, ψ_l, for $\psi_l \in [0, 2\pi]$, is passed undistorted by each filter, while minimizing the total power of the filter output, we may form an estimate of the (squared) amplitudes as

$$\hat{A}_l^2 = \mathrm{E}\left\{|\mathbf{h}_l^H \mathbf{x}(n)|^2\right\} = \mathbf{h}_l^H \mathrm{E}\left\{\mathbf{x}(n)\mathbf{x}(n)^H\right\} \mathbf{h}_l \triangleq \mathbf{h}_l^H \mathbf{R}\mathbf{h}_l,$$ (5.11)

for $l = 1, \ldots, L$. The objective to minimize the total filtered power implies that the filters should be designed such that

$$\mathbf{h}_l = \arg\min_{\mathbf{h}_l} \mathbf{h}_l^H \mathbf{R}\mathbf{h}_l, ,$$ (5.12)

while the requirement to pass the frequency of interest appropriately leads to the constraint

$$\mathbf{h}_l^H \mathbf{z}(\psi_l) = 1,$$ (5.13)

where

$$\mathbf{z}(\psi_l) = \begin{bmatrix} 1 & e^{j\psi_l} & \cdots & e^{j\psi_l(M-1)} \end{bmatrix}^T,$$ (5.14)

i.e., the filter only constrains the frequency of interest to be passed undistorted, while minimizing the power from all other frequencies. We stress that for each frequency of interest, each filter is then only constrained for this single frequency, effectively ignoring the finer structure of the signal. As an alternative, one may instead impose L constraints to each filter, such that

$$\mathbf{h}_l^H \mathbf{Z} = \begin{bmatrix} \underbrace{0 \ \cdots \ 0}_{l-1} & 1 & \underbrace{0 \ \cdots \ 0}_{L-l} \end{bmatrix} \triangleq \mathbf{b}_l,$$ (5.15)

which not only pass the lth frequency undistorted, but also actively places nulls at the frequencies of the other $(L-1)$ frequencies. Using Lagrange multipliers, it is easy to show that the filter solving (5.12), under the constraint (5.13), is found as (see, e.g., [6, 156])

$$\mathbf{h}_l = \frac{\mathbf{R}^{-1}\mathbf{z}(\psi_l)}{\mathbf{z}^H(\psi_l)\mathbf{R}^{-1}\mathbf{z}(\psi_l)},$$ (5.16)

suggesting the classical (power spectrum) Capon amplitude (CCA) estimator

$$\hat{A}_l = \sqrt{\mathbf{h}_l^H \mathbf{R}\mathbf{h}_l} = \left(\mathbf{z}^H(\psi_l)\mathbf{R}^{-1}\mathbf{z}(\psi_l)\right)^{-1/2}$$ (5.17)

for $l = 1, \ldots, L$ under the assumption that $\mathbf{R}$ is invertible, which is the case for $M < \frac{N}{2} + 1$. The estimate in (5.17) is highly structured and allows for an efficient implementation by exploiting the

Gohberg-Semencul formula to find a closed-form expression for $\mathbf{R}^{-1}$ or, alternatively, the generalized Schur algorithm to form an estimate of the Cholesky factor of $\mathbf{R}^{-1}$, in combination with fast FFT-based matrix-vector multiplication techniques (see, e.g., [60, 102, 121] for further details).

Reminiscent to the derivations in Chapter 3, introducing the Lagrange multipliers $\lambda = \begin{bmatrix} \lambda_1 & \cdots & \lambda_L \end{bmatrix}$, the Lagrange dual function for solving (5.12) under the constraint (5.15) can be written as

$$\mathcal{L}(\mathbf{h}_l, \lambda) = \mathbf{h}_l^H \mathbf{R} \mathbf{h}_l - \left(\mathbf{h}_l^H \mathbf{Z} - \mathbf{b}_l \right) \lambda. \tag{5.18}$$

Taking the derivatives with respect to the unknown filter impulse response, $\mathbf{h}_l$, as well as the Lagrange multiplier, we get

$$\nabla \mathcal{L}(\mathbf{h}_l, \lambda) = \begin{bmatrix} \mathbf{R} & -\mathbf{Z} \\ -\mathbf{Z}^H & \mathbf{0} \end{bmatrix} \begin{bmatrix} \mathbf{h}_l \\ \lambda \end{bmatrix} + \begin{bmatrix} \mathbf{0} \\ \mathbf{b}_l \end{bmatrix} \tag{5.19}$$

which, if setting $\nabla \mathcal{L}(\mathbf{h}_l, \lambda) = \mathbf{0}$, yields

$$\lambda = \left(\mathbf{Z}^H \mathbf{R}^{-1} \mathbf{Z} \right)^{-1} \mathbf{b}_l \tag{5.20}$$
$$\mathbf{h}_l = \mathbf{R}^{-1} \mathbf{Z} \lambda. \tag{5.21}$$

If combined, this implies that

$$\mathbf{h}_l = \mathbf{R}^{-1} \mathbf{Z} \left(\mathbf{Z}^H \mathbf{R}^{-1} \mathbf{Z} \right)^{-1} \mathbf{b}_l \tag{5.22}$$

which leads to the multiple constraint Capon amplitude (MCA) estimate given by

$$\hat{A}_l = \sqrt{ \mathbf{b}_l^T \left(\mathbf{Z}^H \mathbf{R}^{-1} \mathbf{Z} \right)^{-1} \mathbf{b}_l } \tag{5.23}$$

for $l = 1, \ldots, L$. We note that using the matrix inversion lemma, the estimator in (5.23) may be formed iteratively over l in a computationally efficient manner using the FFT [79].

As a third alternative, one may use (5.9) to instead form a weighted LS estimate of the unknown amplitude vector [155]

$$\hat{\mathbf{a}} = \left[\sum_{n=0}^{N-M} \mathbf{Z}^H(n) \widehat{\mathbf{Q}}^{-1} \mathbf{Z}(n) \right]^{-1} \left[\sum_{n=0}^{N-M} \mathbf{Z}^H(n) \widehat{\mathbf{Q}}^{-1} \mathbf{x}(n) \right], \tag{5.24}$$

where $\widehat{\mathbf{Q}}$ denotes an estimate of the noise covariance matrix, which is assumed to be invertible. To form such an estimate, introduce the (forward-only) sample covariance matrix estimate

$$\widehat{\mathbf{R}} = \frac{1}{N-M+1} \sum_{n=0}^{N-M} \mathbf{x}(n) \mathbf{x}^H(n) \tag{5.25}$$

which, for large $(N - M + 1)$, converges to $\mathbf{R} = \mathbf{Z} \mathbf{P} \mathbf{Z}^H + \mathbf{Q}$, where

$$\mathbf{P} = \begin{bmatrix} A_1^2 & & 0 \\ & \ddots & \\ 0 & & A_L^2 \end{bmatrix}. \tag{5.26}$$

As shown in [155], for sufficiently large N and M,

$$\widehat{\mathbf{Q}}^{-1}\mathbf{Z}(n) = \widehat{\mathbf{R}}^{-1}\mathbf{Z}\left(\widehat{\mathbf{P}}\mathbf{Z}^H\widehat{\mathbf{Q}}^{-1}\mathbf{Z} + \mathbf{I}\right)\mathbf{D}_n \approx \widehat{\mathbf{R}}^{-1}\mathbf{Z}\mathbf{D}_n\left(\widehat{\mathbf{P}}\mathbf{Z}^H\widehat{\mathbf{Q}}^{-1}\mathbf{Z} + \mathbf{I}\right), \tag{5.27}$$

where $\widehat{\mathbf{P}}$ is some initial estimate of $\mathbf{P}$. Inserting (5.27) in (5.24), noting that $\widehat{\mathbf{P}}\mathbf{Z}^H\widehat{\mathbf{Q}}^{-1}\mathbf{Z} + \mathbf{I}$ will cancel out, yields the extended (amplitude spectrum) Capon amplitude (ECA) estimator

$$\hat{\mathbf{a}} = \left[\sum_{n=0}^{N-M}\mathbf{Z}^H(n)\widehat{\mathbf{R}}^{-1}\mathbf{Z}(n)\right]^{-1}\left[\sum_{n=0}^{N-M}\mathbf{Z}^H(n)\widehat{\mathbf{R}}^{-1}\mathbf{x}(n)\right], \tag{5.28}$$

which, interestingly, does not depend on the initial estimate $\widehat{\mathbf{P}}$. It is worth noting that the ECA estimator can be interpreted as extending the CCA estimator in (5.17) to multiple sinusoids; however, the ECA estimator yields an estimate of the complex-valued amplitudes, whereas the both the CCA and MCA estimators instead estimate the magnitude of the amplitudes.

As a fourth option, one may define L filters for each frequency of interest, forming a matrix filter, $\mathbf{H} = \begin{bmatrix} \mathbf{h}_1 & \cdots & \mathbf{h}_L \end{bmatrix}$, and express the design criteria as

$$\min_{\mathbf{H}} \mathrm{Tr}\left\{\mathbf{H}^H\mathbf{R}\mathbf{H}\right\} \quad \text{subject to} \quad \mathbf{H}^H\mathbf{Z} = \mathbf{I}. \tag{5.29}$$

As shown in (3.39), this filter minimize the sum of the power from all the filters, i.e.,

$$\mathrm{Tr}\left\{\mathbf{H}^H\mathbf{R}\mathbf{H}\right\} = \sum_{l=1}^{L}\mathbf{h}_l^H\mathbf{R}\mathbf{h}_l, \tag{5.30}$$

whereas the constraint in (5.29) ensures that each of the L filters, for each frequency of interest, passes a single sinusoid undistorted while nulling all the others. The filter minimization (5.29) is given by (see, e.g., [6, 156])

$$\mathbf{H} = \mathbf{R}^{-1}\mathbf{Z}(\mathbf{Z}^H\mathbf{R}^{-1}\mathbf{Z})^{-1}, \tag{5.31}$$

which implies that

$$\mathrm{Tr}\left[\mathbf{H}^H\mathbf{R}\mathbf{H}\right] = \mathrm{Tr}\left[(\mathbf{Z}^H\mathbf{R}^{-1}\mathbf{Z})^{-1}\right]. \tag{5.32}$$

Filtering $\mathbf{x}(n)$, defined in (5.9), using this filter yields

$$\mathbf{z}(n) \triangleq \mathbf{H}^H\mathbf{x}(n) = \mathbf{D}_n\mathbf{a} + \mathbf{H}^H\mathbf{e}(n) \triangleq \mathbf{D}_n\mathbf{a} + \mathbf{w}(n), \tag{5.33}$$

implying that the noise covariance matrix, $\widehat{\mathbf{Q}}$, in (5.24) can be formed using an APES-like estimate formed as [155]

$$\widehat{\mathbf{Q}} = \widehat{\mathbf{R}} - \sum_{l=1}^{L}\mathbf{s}_l\mathbf{s}_l^H, \tag{5.34}$$

where

$$\mathbf{s}_l = \frac{1}{N-M+1} \sum_{n=0}^{N-M} \mathbf{x}(n) e^{-j\psi_l n}. \tag{5.35}$$

Inserting (5.34) in (5.24) yields the extended APES amplitude (EAA) estimator. As suggested in [155], an alternative APES-based amplitude estimator may also be formed by instead exploiting the matched-filterbank (MAFI) approach. From (5.33), we note that

$$z_l(n) = a_l e^{j\psi_l n} + w_l(n), \tag{5.36}$$

where $z_l(n)$ and $w_l(n)$ denote the lth element of $\mathbf{z}(n)$ and $\mathbf{w}(n)$, respectively, will only depend on one another via the correlation between $w_l(n)$ and $w_p(n)$, for $l \neq p$. Ignoring this correlation, one may form a least-squares estimate of the complex amplitudes as

$$\hat{a}_l = \frac{1}{N-M+1} \sum_{n=0}^{N-M} z_l(n) e^{-j\psi_l n}. \tag{5.37}$$

The resulting MAFI amplitude estimator offers both reliable and robust amplitude estimates, often being preferable to the other Capon- and APES-based estimators for sinusoids in colored noise [155]. It is worth remarking that all the presented covariance-based amplitude estimators strongly depend on the quality of the covariance matrix estimate. As is well known, the true covariance matrix should be a Toeplitz matrix, whereas the estimate given by (5.25) will, in general, not be Toeplitz. Thus, one could imagine that it would be preferable to instead form the estimate such that the inherent Toeplitz structure is imposed on the estimate. In general, this is not the case, and such estimators will commonly exhibit poor spectral resolution as compared to the outer-product estimate given by (5.25). Instead, it is often better to only require that the covariance matrix estimate $\tilde{\mathbf{R}}$ is persymmetric, i.e., being symmetric about the cross-diagonal[1], such that

$$\tilde{\mathbf{R}} = \mathbf{J}\tilde{\mathbf{R}}^T\mathbf{J} \tag{5.38}$$

where $\mathbf{J}$ is the $M \times M$ exchange matrix

$$\mathbf{J} = \begin{bmatrix} 0 & & 1 \\ & \cdot^{\cdot^{\cdot}} & \\ 1 & & 0 \end{bmatrix} \tag{5.39}$$

As was shown in [83], one often obtain improved spectral estimates by using the persymmetric forward-backward averaged (FB) sample covariance matrix estimate

$$\tilde{\mathbf{R}} = \frac{1}{2}\left(\hat{\mathbf{R}} + \mathbf{J}\hat{\mathbf{R}}^T\mathbf{J}\right) \tag{5.40}$$

[1]This as every Toeplitz matrix is persymmetric, while every persymmetric matrix is not Toeplitz.

where $\widehat{\mathbf{R}}$ is the forward-only (F) sample covariance matrix estimate given by (5.25). This is also the case for the here discussed covariance-based amplitude estimators, which preferably should be formed with an estimate of $\mathbf{R}$ being computed using (5.40) instead of (5.25). The only exceptions are the EAA and MAFI algorithms, which, similarly to the F- and FB-APES versions (see also [106]), require some further attention. Reminiscent to the FB-APES algorithm, one then needs to take the structure of the backward data vector into account, estimating $\widehat{\mathbf{Q}}$ as

$$\widehat{\mathbf{Q}} = \widetilde{\mathbf{R}} - \sum_{l=1}^{L} \mathbf{S}_l \mathbf{S}_l^H, \tag{5.41}$$

where

$$\mathbf{S}_l = \frac{1}{\sqrt{2}} \begin{bmatrix} \mathbf{s}_l & \check{\mathbf{s}}_l \end{bmatrix} \tag{5.42}$$

$$\check{\mathbf{s}}_l = \frac{1}{N-M+1} \sum_{n=0}^{N-M} \check{\mathbf{x}}(n) e^{-j\psi_l n}. \tag{5.43}$$

with $\check{\mathbf{x}}(n)$ denoting the backward data vector

$$\check{\mathbf{x}}(n) = \begin{bmatrix} x^*(N-n-1) & x^*(N-n-2) & \cdots & x^*(N-n-M) \end{bmatrix}^T, \tag{5.44}$$

which is defined for $n = 0, 1, \ldots, N - M$.

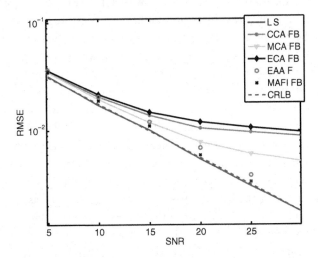

Figure 5.1: RMSE of the discussed amplitude estimators as a function of the local SNR for $N = 160$ and $M = 40$.

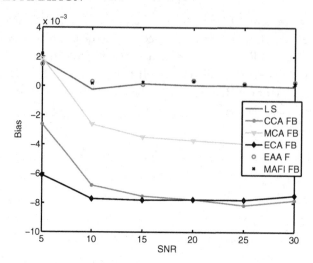

Figure 5.2: Bias of the discussed amplitude estimators as a function of the local SNR, with $N = 160$ and $M = 40$.

5.4 SOME RESULTS AND DISCUSSION

The performance of the discussed estimators, and in particular the data-dependent Capon- and APES-based estimators, will depend on the examined data set. However, some general conclusions may be drawn from numerical studies. For cases with sufficient data lengths, containing well spaced sinusoidal components corrupted by white noise, the LS estimators in (5.4) and (5.6) are often preferable due to their numerical simplicities. For this case, the LS estimator in (5.4) is the maximum likelihood (ML) estimator of the complex amplitudes, whereas (5.6) will only be approximately ML, requiring the spectral lines to be well spaced. For the more realistic colored noise case, the LS estimator in (5.4) will *asymptotically* achieve the same performance[2] as the ML estimator [153]. Similarly, the Capon- and APES-based estimators will both be *asymptotically efficient*. However, for finite data lengths, the estimators have quite different properties. In particular, it was shown in [154] and [106] that the Capon estimator will be biased downwards, whereas the APES estimator will be unbiased (within a second-order approximation). In general, the APES spectral estimator will also have better performance than the Capon estimator, although at the price of a somewhat lower spectral resolution [82]. Examining the here discussed Capon- and APES-based amplitude estimators, one may similarly conclude that the F-EAA and the MAFI estimators will have preferable performance as compared to the Capon-based estimators, which will exhibit a bias downwards. As noted above, the FB-versions are in general preferable, except for the FB-EAA, which suffers from occasional numerical instabilities, making it unreliable. In terms of complexity, the MAFI estimator has a

---

[2]One way to interpret this result is to note that the sinusoidal components have a zero-width support, on which the additive noise can be well modelled as having a flat spectrum.

lower computational cost than the EAA estimator, as well as offering somewhat better performance, suggesting that the former should be preferred in most cases.

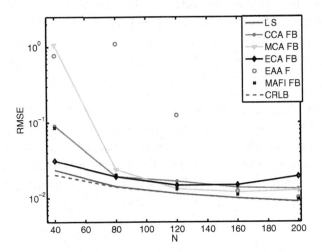

Figure 5.3: RMSE of the discussed amplitude estimators as a function of the data length, using $M = \lfloor N/4 \rfloor$.

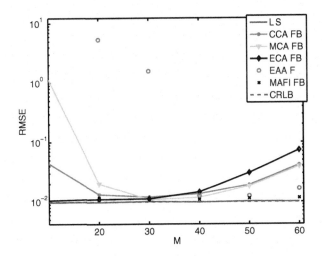

Figure 5.4: RMSE of the discussed amplitude estimators as a function of the filter length, $N = 160$.

Using Monte Carlo simulations, it is easy to confirm the above conclusions on synthetic data. For simplicity, we will here only examine the case of four complex sinusoids corrupted by white Gaussian noise, examining how the performance of the estimators depend on the selected data and filter lengths, as well as the local SNR, defined as A_i^2/σ^2 (see also Section 1.8). The simulation set was selected (arbitrarily) to consist of the absolute frequencies $\psi = 2\pi f$, with $f = 0.12, 0.24, 0.36$ and 0.48, with amplitudes 1, 0.7, 0.5 and 0.3, and with uniformly distributed random phase that is randomized in each iteration. Figures 5.1, 5.3, and 5.4 show the RMSE for the forward-backward estimators (except for EAA) as obtained from 1000 Monte Carlo simulations of the amplitude estimates for the first sinusoidal component as a function of the local SNR, N, and M. As expected, being the ML estimate, the LS estimator in (5.4) offers preferable performance, but the MAFI estimators can be seen to achieve almost the same performance. From Figure 5.1, one may note the puzzling fact that the Capon-based estimators seems to have only a minor performance improvement with increasing SNR. However, this should be expected as the Capon-based estimators will be biased *downwards*, whereas the APES-based estimators will be unbiased [154]. This can also be seen in Figure 5.2, showing the bias of the estimators. Finally, similar to result found in the literature for the Capon- and APES spectral estimators, the figures also show that about $M = \lfloor N/4 \rfloor$ seems to be a reasonable choice for the filter length, although this obviously will depend on the examined data to some extent.

APPENDIX A

The Analytic Signal

As we have seen, the signal models used throughout this book have been complex. We will now show how to map a real signal to a complex one using the Hilbert transform (for an introduction to the Hilbert transform, see [67]). We will now introduce the Hilbert transform and define the so-called analytic signal. Let $x_r(n)$ be a discrete real signal. The discrete Hilbert transform, $\mathcal{H}\{\cdot\}$, of this, denoted $x_i(n)$, is then defined as (see e.g. [125])

$$x_i(n) = \mathcal{H}\{x_r(n)\} = \sum_{m=-\infty}^{\infty} h(m)x_r(n-m). \tag{A.1}$$

where $h(n)$ is the impulse response of the discrete Hilbert transform given by

$$h(n) = \begin{cases} \frac{2\sin^2(\pi n/2)}{\pi n}, & n \neq 0 \\ 0, & n = 0 \end{cases}. \tag{A.2}$$

A useful way of looking at the Hilbert transform, and perhaps a more intuitive definition, is in the frequency domain:

$$X_i(\omega) = H(\omega)X_r(\omega), \quad \text{with} \quad H(\omega) = \begin{cases} -j, & \text{for } 0 < \omega < \pi \\ j, & \text{for } -\pi < \omega < 0 \end{cases}, \tag{A.3}$$

where $X_i(\omega)$ and $X_r(\omega)$ are the Fourier transforms of $x_i(n)$ and $x_r(n)$, respectively, and $H(\omega)$ is the Fourier transform of $h(n)$. The so-called discrete-time analytic signal is then defined as

$$x_c(n) = x_r(n) + jx_i(n). \tag{A.4}$$

It is easy to see that $x_r(n) = \text{Re}\{x_c(n)\}$. By taking the Fourier transform of this expression, we get

$$X_c(\omega) = X_r(\omega) + jX_i(\omega) \tag{A.5}$$
$$= X_r(\omega) + jH(\omega)X_r(\omega), \tag{A.6}$$

and using (A.3), we obtain that the analytic signal has the following spectrum:

$$X_c(\omega) = \begin{cases} 2X_r(\omega), & \text{for } 0 < \omega < \pi \\ 0, & \text{for } -\pi < \omega < 0 \end{cases}. \tag{A.7}$$

It can be seen that the Fourier transform of the analytic signal in (A.4) is zero for negative frequencies and equal to the spectrum of $x_r(n)$ times two for positive frequencies. In other words, a real signal contains complex sinusoids in complex conjugate pairs, while the analytic signal only contains the positive frequencies. Thereby, we have removed half of the sinusoidal components. Since $X_c(\omega)$ is only nonzero in half its spectrum, it is possible to down-sample it by a factor of two, i.e.,

$$x(n) = x_c(2n) \tag{A.8}$$

without loss, assuming that the real signal was obtain by appropriate sampling so that no significant signal energy is present in $X(\pi)$ and $X(0)$. This also leads to a reduction in computational complexity as signal vectors are now only half the length, although as a result the algebra is complex. It is not only advisable to down-sample the signal by a factor of two for complexity reasons, it may also be necessary to avoid rank deficient covariance matrices. If we have N_r real samples of $x_r(n)$, then we have $N = N_r/2$ complex samples $x(N)$ assuming that N_r is even. Consequently, the sampling frequency associated with the analytic signal is $f_c = f_r/2$, where f_r is the sampling frequency at which the real signal $x_r(n)$ has been obtained.

Caveat emptor! Solving estimation problems using the real and analytic signals do not always yield identical estimates or estimators. If there exists considerable signal energy close to 0 and π in $X_r(\omega)$ (relative to N), there will be interaction effects between the negative and positive sides of the spectrum, and by using the analytic signal, this will be ignored. For example, consider a real signal consisting of a complex sinusoid having a low frequency and its complex conjugate. Due to finite window lengths, their side- and main-lobes will overlap and cause the peaks of the spectrum to shift. If the frequency is estimated from the spectrum of the analytic signal, the estimate will be biased due to this interaction. As the number of samples N is increased, these effects will be lessened and we are able to estimate frequencies closer to DC, so under the condition that the number of samples is high, and that usually only audio contents between 20 and 20 kHz is present (or can be filtered out without loss), we should be safe. In many estimators, it is possible to account for the interaction, but in most cases, it is simply not worth it in terms of the added computational burden.

The question remains how to actually compute the analytic signal. The impulse response of the Hilbert transform filter in (A.2) is infinite and cannot be implemented directly. Fortunately, it can be implemented efficiently using the fast Fourier transform (FFT) such that the implementation satisfies that a) the real value of the analytic signal is equal to the original real sequence, and b) the real and imaginary parts of the analytic signal are orthogonal (see [47] for an improved method). More specifically, the down-sampled analytic signal $x(n)$ can be obtained from N_r real samples $x_r(n)$ as follows [112]:

- The N_r-point discrete Fourier transform $X_r(k)$ of $x_r(n)$ is computed using the FFT, i.e.,

$$X_r(k) = \sum_{n=0}^{N-1} x_r(n)e^{-j2\pi\frac{k}{N}n}, \tag{A.9}$$

for $k = 0, \ldots, N_r - 1$.

- From this the $N_r/2$ point discrete Fourier transform of the down-sampled analytic signal $X(k)$ is formed as

$$X(k) = \begin{cases} X_r(0) + X(\frac{N_r}{2}) & \text{for} \quad k = 0 \\ 2X_r(k) & \text{for} \quad 1 \leq k \leq \frac{N_r}{2} - 1 \end{cases}. \tag{A.10}$$

- Finally, we calculate the time-domain down-sampled analytic signal $x(n)$ using an $N_r/2$ point discrete Fourier transform scaled by $1/2$, i.e.,

$$x(n) = \frac{1}{N_r} \sum_{k=0}^{N_r/2-1} X(k) e^{j2\pi \frac{k}{N_r/2} n}, \tag{A.11}$$

for $n = 0, \ldots, N_r/2 - 1$.

The down-sampled analytic signal $x(n)$ can now be used for estimation purposes based on the complex signal models preferred in this book. To recover the original real signal $x_r(n)$ from $x(n)$ or a model thereof, the following procedure can be used:

- First, calculate the $N_r/2$-point discrete Fourier transform of $x(n)$, i.e.,

$$X(k) = \sum_{n=0}^{N_r/2-1} x(n) e^{-j2\pi \frac{k}{N_r/2} n}, \tag{A.12}$$

for $k = 0, \ldots, N_r/2 - 1$.

- Zero-pad this transform to obtain the N_r-point discrete Fourier transform of $x_c(n)$, i.e.,

$$X_c(k) = \begin{cases} X(k) & \text{for} \quad 0 \leq k \leq \frac{N_r}{2} - 1 \\ 0 & \text{for} \quad \frac{N_r}{2} \leq k \leq N_r - 1 \end{cases}. \tag{A.13}$$

- Calculate the analytic signal using the N_r-point inverse Fourier transform, i.e.,

$$x_c(n) = \frac{2}{N_r} \sum_{k=0}^{N_r-1} X_c(k) e^{j2\pi \frac{k}{N_r} n}. \tag{A.14}$$

- From $x_c(n)$ obtain the corresponding real signal as $x_r(n) = \text{Re}\{x_c(n)\}$.

At this point it should be stressed that this particular procedure, although simple, intuitive and computationally efficient, is not without its problems, aside from the problems generally associated with using the ideal analytic signal mentioned earlier. From filter design theory, it is well-known that direct manipulation of the coefficients of a discrete Fourier transform is a problematic, yet tempting, way of designing digital filters. It is sometimes referred to as the frequency sampling design method. The resulting filters may have some poor characteristics such as excessive ripples in between frequency points and this is also the case for this Hilbert transform implementation. Furthermore, aliasing phenomena may also occcur depending on the particular processing taking place. Nonetheless, it is still often worthwhile to use the analytic signal as computed here due to the savings in computation time. Finally, we note that it is possible to formulate a time-recursive updating of the analytic signal [80].

Bibliography

[1] S. S. Abeysekera, "Multiple pitch estimation of poly-phonic audio signals in a frequency-lag domain using the bispectrum," in *Proc. IEEE Int. Symp. Circuits and Systems*, vol. 14(4), 2004, pp. 469–472. DOI: 10.1109/ISCAS.2004.1328785 2

[2] H. Akaike, "A new look at the statistical model identification," *IEEE Trans. Autom. Control*, vol. 19, pp. 716–723, 1974. DOI: 10.1109/TAC.1974.1100705 38

[3] American National Standards Institute (ANSI), "American National Standard Acoustical Terminology," New York, 1994. 1

[4] American Standards Association (ASA), "Acoustical Terminology, SI, 1-1960," New York, 1960. 1, 5

[5] T. W. Anderson, "Asymptotic theory for principal component analysis," *Ann. Math. Stat.*, vol. 34, pp. 122–148, 1963. DOI: 10.1214/aoms/1177704248 87

[6] A. Antoniou and W.-S. Lu, *Practical Optimization: Algorithms and Engineering Applications*. Springer Verlag, 2007. 31, 62, 113, 115

[7] T. M. Apostol, *Mathematical Analysis*, 2nd ed. Addison-Wesley, 1974. 68

[8] R. Badeau, B. David, and G. Richard, "Fast approximated power iteration subspace tracking," *IEEE Trans. Signal Process.*, vol. 53, no. 8, pp. 2931–2941, Aug. 2005. DOI: 10.1109/TSP.2005.850378 99

[9] R. Badeau, B. David, and G. Richard, "A new perturbation analysis for signal enumeration in rotational invariance techniques," *IEEE Trans. Signal Process.*, vol. 54, no. 2, pp. 450–458, Feb. 2006. DOI: 10.1109/TSP.2005.861899 102, 103

[10] R. Badeau, G. Richard, and B. David, "Sliding window adaptive SVD algorithms," *IEEE Trans. Signal Process.*, vol. 52, no. 1, pp. 1–10, Jan 2004. DOI: 10.1109/TSP.2003.820069 99

[11] A. J. Barabell, "Improving the resolution performance of eigenstructure-based direction-finding algorithms," in *Proc. IEEE Int. Conf. Acoust., Speech, Signal Processing*, 1983, pp. 336–339. DOI: 10.1109/ICASSP.1983.1172124 83

[12] G. Bienvenu, "Influence of the spatial coherence of the background noise on high resolution passive methods," in *Proc. IEEE Int. Conf. Acoust., Speech, Signal Processing*, 1979, pp. 306–309. DOI: 10.1109/ICASSP.1979.1170720 81, 83

[13] G. Bienvenu and L. Kopp, "Optimality of high resolution array processing using the eigen-system approach," *IEEE Trans. Acoust., Speech, Signal Process.*, vol. 31(5), pp. 1235–1248, Oct. 1983. DOI: 10.1109/TASSP.1983.1164185 87

[14] S. Boyd and L. Vandenberghe, *Convex Optimization.*   Cambridge University Press, 2004. 37, 62, 101

[15] A. T. Cemgil, "Bayesian music transcription," Ph.D. dissertation, Nijmegen University, 2004. 3, 5, 15

[16] A. T. Cemgil, H. J. Kappen, and D. Barber, "A generative model for music transcription," *IEEE Transactions on Audio, Speech, and Language Processing*, vol. 14, no. 2, pp. 679–694, March 2006. DOI: 10.1109/TSA.2005.852985 3

[17] K. W. Chan and H. C. So, "Accurate frequency estimation for real harmonic sinusoids," *IEEE Signal Process. Lett.*, vol. 11(7), pp. 609–612, July 2004. DOI: 10.1109/LSP.2004.830115 49

[18] D. Chazan, Y. Stettiner, and D. Malah, "Optimal multi-pitch estimation using the em al-gorithm for co-channel speech separation," in *Proc. IEEE Int. Conf. Acoust., Speech, Signal Processing*, vol. 2, 27–30 April 1993, pp. 728–731. DOI: 10.1109/ICASSP.1993.319415 3, 47, 60

[19] M. G. Christensen, "Estimation and modeling problems in parametric audio coding," Ph.D. dissertation, Aalborg University, 2005. 4

[20] M. G. Christensen, "On perceptual distortion measures and parametric modeling," in *Proc. Acoustics'08 Paris*, 2008. DOI: 10.1121/1.2935505 3

[21] M. G. Christensen, A. Jakobsson, and S. H. Jensen, "Multi-pitch estimation using har-monic MUSIC," in *Rec. Asilomar Conf. Signals, Systems, and Computers*, 2006, pp. 521–525. DOI: 10.1109/ACSSC.2006.354802 xiii, 81, 98

[22] M. G. Christensen, A. Jakobsson, and S. H. Jensen, "Fundamental frequency estimation using the shift-invariance property," in *Rec. Asilomar Conf. Signals, Systems, and Computers*, 2007, pp. 631–635. DOI: 10.1109/ACSSC.2007.4487290 xiii, 107, 108, 109

[23] M. G. Christensen, A. Jakobsson, and S. H. Jensen, "Joint high-resolution fundamental fre-quency and order estimation," *IEEE Trans. Audio, Speech, and Language Process.*, vol. 15(5), pp. 1635–1644, July 2007. DOI: 10.1109/TASL.2007.899267 xiii, 16, 21, 40, 51, 55, 81, 98, 108, 109

[24] M. G. Christensen, A. Jakobsson, and S. H. Jensen, "Sinusoidal order estimation using the subspace orthogonality and shift-invariance properties," in *Rec. Asilomar Conf. Signals, Systems, and Computers*, 2007, pp. 651–655. DOI: 10.1109/ACSSC.2007.4487294 xiii

[25] M. G. Christensen, A. Jakobsson, and S. H. Jensen, "Sinusoidal order estimation based on angles between subspaces," Feb. 2009, unpublished manuscript. xiii

[26] M. G. Christensen, J. H. Jensen, A. Jakobsson, and S. H. Jensen, "On optimal filter designs for fundamental frequency estimation," *IEEE Signal Process. Lett.*, vol. 15, pp. 745–748, 2008. DOI: 10.1109/LSP.2008.2003987 xiii, 69, 73

[27] M. G. Christensen, J. H. Jensen, A. Jakobsson, and S. H. Jensen, "Joint fundamental frequency and order estimation using optimal filtering," Feb. 2009, unpublished manuscript. xiii, 76, 80

[28] M. G. Christensen and S. H. Jensen, "New results in rate-distortion optimized parametric audio coding," in *Proc. 120th AES Convention*, 2006, preprint 6808. 4

[29] M. G. Christensen and S. H. Jensen, "On perceptual distortion minimization and nonlinear least-squares frequency estimation," *IEEE Trans. Audio, Speech, and Language Process.*, vol. 14(1), pp. 99–109, Jan. 2006. DOI: 10.1109/TSA.2005.860347 3

[30] M. G. Christensen and S. H. Jensen, "Variable order harmonic sinusoidal parameter estimation for speech and audio signals," in *Rec. Asilomar Conf. Signals, Systems, and Computers*, 2006, pp. 1126–1130. DOI: 10.1109/ACSSC.2006.354929 xiii, 43, 55

[31] M. G. Christensen, S. H. Jensen, S. V. Andersen, and A. Jakobsson, "Subspace-based fundamental frequency estimation," in *Proc. European Signal Processing Conf.*, 2004, pp. 637–640. xiii

[32] M. G. Christensen, P. Stoica, A. Jakobsson, and S. H. Jensen, "The multi-pitch estimation problem: Some new solutions," in *Proc. IEEE Int. Conf. Acoust., Speech, Signal Processing*, vol. 3, 2007, pp. 1221–1224. DOI: 10.1109/ICASSP.2007.367063 xiii

[33] M. G. Christensen, P. Stoica, A. Jakobsson, and S. H. Jensen, "Multi-pitch estimation," *Elsevier Signal Processing*, vol. 88(4), pp. 972–983, Apr. 2008. DOI: 10.1016/j.sigpro.2007.10.014 xiii, 31, 44, 55, 64, 80, 81, 98, 109

[34] M. G. Christensen, P. Vera-Candeas, S. D. Somasundaram, and A. Jakobsson, "Robust subspace-based fundamental frequency estimation," in *Proc. IEEE Int. Conf. Acoust., Speech, Signal Processing*, 2008, pp. 101–104. DOI: 10.1109/ICASSP.2008.4517556 xiii, 16, 17, 18, 81

[35] D. Clark, A. T. Cemgil, P. Peeling, and S. Godsill, "Multi-object tracking of sinusoidal components in audio with the gaussian mixture probability hypothesis density filter," in *Proc. IEEE Workshop on Applications of Signal Processing to Audio and Acoustics*, 21–24 Oct. 2007, pp. 339–342. DOI: 10.1109/ASPAA.2007.4393009 15

[36] P. Comon and G. Golub, "Tracking a few extreme singular values and vectors in signal processing," *Proc. IEEE*, vol. 78, no. 8, pp. 1327–1343, Aug. 1990. DOI: 10.1109/5.58320 98

[37] T. Dau, D. Püschel, and A. Kohlrausch, "A quantitative model of the effective signal processing in the auditory system. i. model structure," *J. Acoust. Soc. Am.*, vol. 99(6), pp. 3615–3622, June 1996. DOI: 10.1121/1.414959 3

[38] T. Dau, D. Püschel, and A. Kohlrausch, "A quantitative model of the effective signal processing in the auditory system. ii. simulations and measurements," *J. Acoust. Soc. Am.*, vol. 99(6), pp. 3623–3631, June 1996. DOI: 10.1121/1.414960 3

[39] B. David and R. Badeau, "Fast sequential LS estimation for sinusoidal modeling and decomposition of audio signals," in *Proc. IEEE Workshop on Appl. of Signal Process. to Aud. and Acoust.*, 2007, pp. 211–214. DOI: 10.1109/ASPAA.2007.4392992 30

[40] M. Davy, S. Godsill, and J. Idier, "Bayesian analysis of western tonal music," *J. Acoust. Soc. Am.*, vol. 119(4), pp. 2498–2517, Apr. 2006. DOI: 10.1121/1.2168548 3

[41] A. de Cheveigné and H. Kawahara, "YIN, a fundamental frequency estimator for speech and music," *J. Acoust. Soc. Am.*, vol. 111(4), pp. 1917–1930, Apr. 2002. DOI: 10.1121/1.1458024 2, 23

[42] P. de la Cuadra, A. Master, and C. Sapp, "Efficient pitch detection techniques for interactive music," in *Proc. Int. Computer Music Conf.*, 2001. 2

[43] P. M. Djuric, "A model selection rule for sinusoids in white gaussian noise," *IEEE Trans. Signal Process.*, vol. 44, no. 7, pp. 1744–1751, July 1996. DOI: 10.1109/78.510621 38, 39

[44] P. M. Djuric, "Asymptotic MAP criteria for model selection," *IEEE Trans. Signal Process.*, vol. 46, pp. 2726–2735, Oct. 1998. DOI: 10.1109/78.720374 38, 39, 41, 109

[45] X. Doukopoulos and G. Moustakides, "Fast and stable subspace tracking," *IEEE Trans. Signal Process.*, vol. 56, no. 4, pp. 1452–1465, April 2008. DOI: 10.1109/TSP.2007.909335 99

[46] C. Dubois and M. Davy, "Joint detection and tracking of time-varying harmonic components: A flexible bayesian approach," *IEEE Transactions on Audio, Speech, and Language Processing*, vol. 15, no. 4, pp. 1283–1295, May 2007. DOI: 10.1109/TASL.2007.894522 15

[47] M. Elfataoui and G. Mirchandani, "A frequency-domain method for generation of discrete-time analytic signals," *IEEE Trans. Signal Process.*, vol. 54, no. 9, pp. 3343–3352, Sept. 2006. DOI: 10.1109/TSP.2006.879302 122

[48] V. Emiya, R. Badeau, and B. David, "Multipitch estimation of quasi-harmonic sounds in colored noise," in *Proc. Int. Conf. Digital Audio Effects*, 2007. 29

[49] V. Emiya, B. David, and R. Badeau, "A parametric method for pitch estimation of piano tones," in *Proc. IEEE Int. Conf. Acoust., Speech, Signal Processing*, vol. 1, 2007, pp. 249–252. DOI: 10.1109/ICASSP.2007.366663 2, 18

[50] A. Eriksson, P. Stoica, and T. Söderström, "Asymptotical analysis of MUSIC and ESPRIT frequency estimates," in *Proc. IEEE Int. Conf. Acoust., Speech, Signal Processing*, vol. 4, 1993, pp. 556–559. DOI: 10.1109/ICASSP.1993.319718 84

[51] M. Feder and E. Weinstein, "Parameter estimation of superimposed signals using the EM algorithm," *IEEE Trans. Acoust., Speech, Signal Process.*, vol. 36(4), pp. 477–489, Apr. 1988. DOI: 10.1109/29.1552 44, 47

[52] P. Fernandez-Cid and F. J. Casajus-Quiros, "Multi-pitch estimation for polyphonic musical signals," in *Proc. IEEE Int. Conf. Acoust., Speech, Signal Processing*, vol. 6, 1998, pp. 12–15. DOI: 10.1109/ICASSP.1998.679645 2

[53] H. Fletcher, "Normal vibration frequencies of a stiff piano string," in *J. Acoust. Soc. Amer.*, vol. 36(1), 1962. DOI: 10.1121/1.1919187 17, 18

[54] N. H. Fletcher and T. D. Rossing, *The Physics of Musical Instruments*, 2nd ed. Springer, 1998. 17, 99

[55] J.-J. Fuchs, "Estimating the number of sinusoids in additive white noise," *IEEE Trans. Acoust., Speech, Signal Process.*, vol. 36(12), pp. 1846–1853, 1988. DOI: 10.1109/29.9029 42

[56] J.-J. Fuchs, "Estimating the number of signals in the presence of correlated sensor noise," *IEEE Trans. Acoust., Speech, Signal Process.*, vol. 40(5), pp. 1053–1061, July 1992. DOI: 10.1109/78.134468 42

[57] E. B. George and M. J. T. Smith, "Analysis-by-synthesis/overlap-add sinusoidal modeling applied to the analysis-synthesis of musical tones," *J. Audio Eng. Soc.*, vol. 40(6), pp. 497–516, 1992. 5

[58] E. B. George and M. J. T. Smith, "Speech analysis/synthesis and modification using an analysis-by-synthesis/overlap-add sinusoidal model," *IEEE Trans. Speech Audio Process.*, vol. 5(5), pp. 389–406, Sept. 1997. DOI: 10.1109/89.622558 5, 18

[59] D. Giacobello, M. G. Christensen, J. Dahl, S. H. Jensen, and M. Moonen, "Sparse linear predictors for speech processing," in *Proc. Interspeech*, 2008. 15

[60] G.-O. Glentis, "A fast algorithm for APES and Capon spectral estimation," *IEEE Trans. Signal Processing*, vol. 56, no. 9, pp. 4207–4220, 2008. DOI: 10.1109/TSP.2008.925940 114

[61] S. Godsill and M. Davy, "Bayesian harmonic models for musical pitch estimation and analysis," in *Proc. IEEE Int. Conf. Acoust., Speech, Signal Processing*, vol. 2, 2002, pp. 1769–1772. DOI: 10.1109/ICASSP.2002.1006106 3, 15, 19

[62] S. Godsill and M. Davy, "Bayesian computational models for inharmonicity in musical instruments," in *Proc. IEEE Workshop on Appl. of Signal Process. to Aud. and Acoust.*, 2005, pp. 283–286. DOI: 10.1109/ASPAA.2005.1540225 3, 18

[63] B. Gold and L. Rabiner, "Parallel processing techniques for estimating pitch periods of speech in the time domain," *J. Acoust. Soc. Am.*, vol. 46, pp. 442–448, Aug. 1969. DOI: 10.1121/1.1911709 2

[64] G. H. Golub and C. F. V. Loan, *Matrix Computations*, 3rd ed.   The Johns Hopkins University Press, 1996. 85, 89

[65] R. M. Gray, "Toeplitz and circulant matrices: A review," *Foundations and Trends in Communications and Information Theory*, vol. 2(3), pp. 155–239, 2006. DOI: 10.1561/0100000006 68, 70

[66] R. Gribonval and E. Bacry, "Harmonic Decomposition of Audio Signals with Matching Pursuit," *IEEE Trans. Signal Process.*, vol. 51(1), pp. 101–111, Jan. 2003. DOI: 10.1109/TSP.2002.806592 48, 49

[67] S. L. Hahn, *Hilbert Transforms in Signal Processing*.   Artech House, 1996. 121

[68] E. J. Hannan, *Developments in Time Series Analysis*.   Chapman and Hall, 1993, ch. Determining the number of jumps in a spectrum, pp. 127–138. 42

[69] E. J. Hannan and B. Wahlberg, "Convergence rates for inverse toeplitz matrix forms," *J. Multivariate Analysis*, vol. 31, pp. 127–135, 1989. DOI: 10.1016/0047-259X(89)90055-9 68, 69

[70] E. J. Hannan, "Time series analysis," *IEEE Trans. Autom. Control*, vol. 19(6), pp. 706–715, Dec. 1974. DOI: 10.1109/TAC.1974.1100732 32

[71] P. Hansen and S. Jensen, "Fir filter representations of reduced-rank noise reduction," *IEEE Trans. Signal Process.*, vol. 46, no. 6, pp. 1737–1741, 1998. DOI: 10.1109/78.678511 81

[72] P. Hansen and S. Jensen, "Prewhitening for rank-deficient noise in subspace methods for noise reduction," *IEEE Trans. Signal Process.*, vol. 53, no. 10, pp. 3718–3726, 2005. DOI: 10.1109/TSP.2005.855110 85

[73] S. Haykin, *Adaptive Filter Theory*, 3rd ed.   Prentice-Hall, 1996. 74

[74] D. J. Hermes, "Measurement of pitch by subharmonic summation," *J. Acoust. Soc. Am.*, vol. 83(1), pp. 257–264, 1988. DOI: 10.1121/1.396427 32

[75] W. Hess, *Pitch Determination of Speech Signals*.   Springer-Verlag, Berlin, 1983. 2

[76] W. Hess, "Pitch and voicing determination," in *Advances in Speech Signal Processing*, S. Furui and M. M. Sohndi, Eds.   Marcel Dekker, New York, 1992, pp. 3–48. 2

[77] G. H. Händel, "On the history of music," *IEEE Signal Process. Mag.*, p. 13, Mar. 1999. 83

[78] "ISO/IEC 14496-3:2001/AMD2," ISO/IEC, July 2004, parametric Coding for High-Quality Audio. 4

[79] A. Jakobsson, S. R. Alty, and S. Lambotharan, "On the Implementation of the Linearly Constrained Minimum Variance Beamformer," *IEEE Trans. Circuits & Systems, Part II*, vol. 53, pp. 1059–1061, October 2006. DOI: 10.1109/TCSII.2006.882228 114

[80] A. Jakobsson, S. V. Andersen, and S. R. Alty, "Time-updating discrete-time 'analytic' signals," *IEE Electronic Letters*, vol. 40, no. 9, pp. 205–206, February 2004. DOI: 10.1049/el:20040148 124

[81] A. Jakobsson, M. G. Christensen, and S. H. Jensen, "Frequency selective sinusoidal order estimation," *IEE Electronic Letters*, vol. 43(21), pp. 1164–1165, Oct. 2007. DOI: 10.1049/el:20071738 xiii

[82] A. Jakobsson and P. Stoica, "Combining Capon and APES for Estimation of both Amplitude and Frequency of Spectral Lines," *Circuits, Systems, and Signal Processing*, vol. 19, no. 2, pp. 159–169, 2000. DOI: 10.1007/BF01212468 118

[83] M. Jansson and P. Stoica, "Forward-Only and Forward-Backward Sample Covariances – A Comparative Study," *Elsevier Signal Processing*, vol. 77, no. 3, pp. 235–245, 1999. DOI: 10.1016/S0165-1684(99)00037-7 116

[84] J. Jensen and J. H. L. Hansen, "Speech enhancement using a constrained iterative sinusoidal model," *IEEE Trans. Speech Audio Process.*, vol. 9, pp. 731–740, Oct. 2001. DOI: 10.1109/89.952491 4

[85] J. R. Jensen, J. K. Nielsen, M. G. Christensen, J. S. H., and T. Larsen, "On fast implementation of harmonic music for known and unknown model orders," in *Proc. European Signal Processing Conf.*, 2008. xiii, 99

[86] S. H. Jensen, P. C. Hansen, S. D. Hansen, and J. A. Sorensen, "Reduction of broad-band noise in speech by truncated QSVD," *IEEE Trans. Speech Audio Process.*, vol. 3, no. 6, pp. 439–448, 1995. DOI: 10.1109/89.482211 81

[87] L. Kavalieris and E. J. Hannan, "Determining the number of terms in a trigonometric regression," *J. on Time Series Analysis*, vol. 15(6), pp. 613–625, 1994. DOI: 10.1111/j.1467-9892.1994.tb00216.x 42

[88] S. M. Kay, *Modern Spectral Estimation: Theory and Application.* Prentice-Hall, 1988. 21, 50, 52

[89] S. M. Kay, *Fundamentals of Statistical Signal Processing: Estimation Theory.* Prentice-Hall, 1993. 1, 19, 20

[90] S. M. Kay, *Fundamentals of Statistical Signal Processing: Detection Theory.* Prentice-Hall, 1998. 1

[91] K. Kim and G. Shevlyakov, "Why gaussianity?" *IEEE Signal Process. Mag.*, vol. 25, no. 2, pp. 102–113, March 2008. DOI: 10.1109/MSP.2007.913700 15

[92] A. Klapuri, "Multiple fundamental frequency estimation based on harmonicity and spectral smoothness," *IEEE Trans. Speech Audio Process.*, vol. 11(6), pp. 804–816, 2003. DOI: 10.1109/TSA.2003.815516 2, 17, 37

[93] A. Klapuri, "Multipitch analysis of polyphonic music and speech signals using an auditory model," *IEEE Transactions on Audio, Speech, and Language Processing*, vol. 16, no. 2, pp. 255–266, Feb. 2008. DOI: 10.1109/TASL.2007.908129 3

[94] A. Klapuri and M. Davy, Eds., *Signal Processing Methods for Music Transcription.* New York: Springer, 2006. DOI: 10.1007/0-387-32845-9 3

[95] W. B. Kleijn and K. K. Paliwal, Eds., *Speech Coding and Synthesis.* Elsevier Science B.V., 1995. 4

[96] B. Kostek, "Musical instrument classification and duet analysis employing music information retrieval techniques," *Proc. IEEE*, vol. 92, no. 4, pp. 712–729, Apr 2004. DOI: 10.1109/JPROC.2004.825903 4

[97] D. Kundu and S. Nandi, "A note on estimating the fundamental frequency of a periodic function," *Elsevier Signal Processing*, vol. 84, pp. 653–661, 2004. DOI: 10.1016/j.sigpro.2003.11.016

[98] H. Krim and M. Viberg, "Two decades of array signal processing research–the parametric approach," *IEEE Signal Process. Mag.*, July 1996. DOI: 10.1109/79.526899 81, 84

[99] P. Ladefoged and I. Maddieson, *The Sounds of the World's Languages.* Oxford: Blackwell, 1996. 23

[100] M. A. Lagunas, M. E. Santamaria, A. Gasull, and A. Moreno, "Maximum likelihood filters in spectral estimation problems," *Elsevier Signal Processing*, vol. 10(1), pp. 19–34, 1986. DOI: 10.1016/0165-1684(86)90062-9 73

[101] N. M. Laird, A. P. Dempster, and D. B. Rubin, "Maximum likelihood from incomplete data via the EM algorithm," *Ann. Roy. Stat. Soc.*, pp. 1–38, Dec. 1977. 44

[102] E. G. Larsson and P. Stoica, "Fast Implementation of Two-Dimensional APES and Capon Spectral Estimators," *Multidimensional Systems and Signal Processing*, vol. 13, no. 1, pp. 35–54, Jan. 2002. DOI: 10.1023/A:1013891327453 114

[103] J. Lattard, "Influence of inharmonicity on the tuning of a piano - measurements and mathematical simulation," *J. Acoust. Soc. Am.*, vol. 94, pp. 46–53, 1993. DOI: 10.1121/1.407059 18

[104] M. Lavielle and C. Lévy-Leduc, "Semiparametric estimation of the frequency of unknown periodic functions and its application to laser vibrometry signals," *IEEE Trans. Signal Process.*, vol. 53(7), pp. 2306–2314, July 2005. DOI: 10.1109/TSP.2005.849156 2

[105] P. Leveau, E. Vincent, G. Richard, and L. Daudet, "Instrument-specific harmonic atoms for mid-level music representation," *IEEE Trans. Audio, Speech, and Language Process.*, vol. 16(1), pp. 116–128, Jan. 2008. DOI: 10.1109/TASL.2007.910786 2

[106] H. Li, J. Li, and P. Stoica, "Performance analysis of forward-backward matched-filterbank spectral estimators," *IEEE Trans. Signal Process.*, vol. 46, no. 7, pp. 1954–1966, July 1998. DOI: 10.1109/78.700967 80, 117, 118

[107] H. Li, P. Stoica, and J. Li, "Computationally efficient parameter estimation for harmonic sinusoidal signals," *Signal Processing*, vol. 80, pp. 1937–1944, 2000. DOI: 10.1016/S0165-1684(00)00103-1 21, 49, 51, 55

[108] A. P. Liavas and P. A. Regalia, "On the behavior of information theoretic criteria for model order selection," *IEEE Trans. Signal Process.*, vol. 49(8), pp. 1689–1695, Aug. 2001. DOI: 10.1109/78.934138 88

[109] J. Lim, A. Oppenheim, and L. Braida, "Evaluation of an adaptive comb filtering method for enhancing speech degraded by white noise addition," *IEEE Trans. Acoust., Speech, Signal Process.*, vol. 26, no. 4, pp. 354–358, Aug 1978. DOI: 10.1109/TASSP.1978.1163117 58

[110] J. Lindblom, "A sinusoidal voice over packet coder tailored for the frame-erasure channel," *IEEE Trans. Speech Audio Process.*, vol. 13(5), pp. 16–19, 2005. DOI: 10.1109/TSA.2005.851913 4, 5

[111] M. W. Mak, S. Y. Kung, and S. H. Lin, *Biometric Authentication: A Machine Learning Approach.* Prentice Hall, 2004. 45

[112] S. L. Marple, "Computing the discrete-time "analytic" signal via FFT," *IEEE Trans. Signal Process.*, vol. 47, pp. 2600–2603, Sept. 1999. DOI: 10.1109/78.782222 122

[113] R. J. McAulay and T. F. Quatieri, "Speech analysis/synthesis based on a sinusoidal representation," *IEEE Trans. Acoust., Speech, Signal Process.*, vol. 34(4), pp. 744–754, Aug. 1986. DOI: 10.1109/TASSP.1986.1164910 5

[114] R. J. McAulay and T. F. Quatieri, "Speech transformation based on a sinusoidal representation," *IEEE Trans. Acoust., Speech, Signal Process.*, vol. 34, pp. 1449–1464, Dec. 1986. DOI: 10.1109/TASSP.1986.1164910 5

[115] R. J. McAulay and T. F. Quatieri, "Sinusoidal coding," in *Speech Coding and Synthesis*, W. B. Kleijn and K. K. Paliwal, Eds.   Elsevier Science B.V., 1995, ch. 4, pp. 121–174. 4

[116] R. J. McAulay and E. M. Hofstetter, "Barankin bounds on parameter estimation," *IEEE Trans. Inf. Theory*, vol. 17, no. 6, pp. 669–676, Nov. 1971. DOI: 10.1109/TIT.1971.1054719 19

[117] Y. Medan, E. Yair, and D. Chazan, "Super resolution pitch determination of speech signals," *IEEE Trans. Signal Process.*, vol. 39, no. 1, pp. 40–48, Jan. 1991. DOI: 10.1109/78.80763 2

[118] Y. Miao and Y. Hua, "Fast subspace tracking and neural network learning by a novel information criterion," *IEEE Trans. Signal Process.*, vol. 46, no. 7, pp. 1967–1979, July 1998. DOI: 10.1109/78.700968 99

[119] B. C. J. Moore, *An Introduction to the Psychology of Hearing*, 4th ed.   Academic Press, 1997. 3

[120] J. Moorer, "The optimum comb method of pitch period analysis of continuous digitized speech," *IEEE Trans. Acoust., Speech, Signal Process.*, vol. 22, no. 5, pp. 330–338, Oct 1974. DOI: 10.1109/TASSP.1974.1162596 58

[121] B. Musicus, "Fast MLM power spectrum estimation from uniformly spaced correlations," *IEEE Transactions on Acoustics, Speech and Signal Processing*, vol. 33, no. 4, pp. 1333–1335, 1985. DOI: 10.1109/TASSP.1985.1164696 114

[122] A. Nehorai and B. Porat, "Adaptive comb filtering for harmonic signal enhancement," *IEEE Trans. Acoust., Speech, Signal Process.*, vol. 34(5), pp. 1124–1138, Oct. 1986. DOI: 10.1109/TASSP.1986.1164952 21, 58, 60, 80

[123] A. M. Noll, "Cepstrum pitch determination," *J. Acoust. Soc. Am.*, vol. 41(2), pp. 293–309, 1967. DOI: 10.1121/1.1910339 2

[124] M. Noll, "Pitch determination of human speech by harmonic product spectrum, the harmonic sum, and a maximum likelihood estimate," in *Proc. Symposium on Computer Processing Communications*, 1969, pp. 779–797. 2, 32

[125] A. V. Oppenheim and R. W. Schafer, *Discrete-Time Signal Processing*, 1st ed.   Prentice-Hall, 1989. 121

[126] S. J. Orfanidis, *Introduction to Signal Processing*.   Prentice-Hall International, Inc., 1996. 58

[127] J.-M. Papy, L. De Lathauwer, and S. van Huffel, "A shift invariance-based order-selection technique for exponential data modeling," *IEEE Signal Process. Lett.*, vol. 14(7), pp. 473–476, 2007. DOI: 10.1109/LSP.2006.891324 102

[128] V. F. Pisarenko, "The retrieval of harmonics from a covariance function," *Geophys. J. Roy. Astron. Soc.*, vol. 33, pp. 347–366, 1973. DOI: 10.1111/j.1365-246X.1973.tb03424.x 81

[129] C. J. Plack, A. J. O. R. R. Fay, and A. N. Popper, Eds., *Pitch: Neural Coding and Perception*, ser. Springer Handbook of Auditory Research. Springer Science, 2005. 1, 3, 4

[130] G. E. Poliner and D. P. W. Ellis, "A discriminative model for polyphonic piano transcription," *EURASIP J. on Advances in Signal Processing*, 2009. 2

[131] G. E. Poliner, D. P. W. Ellis, A. F. Ehmann, E. Gomez, S. Streich, and B. Ong, "Melody transcription from music audio: Approaches and evaluation," *IEEE Transactions on Audio, Speech, and Language Processing*, vol. 15, no. 4, pp. 1247–1256, May 2007. DOI: 10.1109/TASL.2006.889797 2

[132] P. Prandoni, M. M. Goodwin, and M. Vetterli, "Optimal time segmentation for signal modeling and compression," in *Proc. IEEE Int. Conf. Acoust., Speech, Signal Processing*, 1997, pp. 2029–2032. DOI: 10.1109/ICASSP.1997.599343 15

[133] P. Prandoni and M. Vetterli, "R/D optimal linear prediction," *IEEE Trans. Speech Audio Process.*, pp. 646–655, 8(6) 2000. DOI: 10.1109/89.876298 15

[134] H. Purnhagen and N. Meine, "HILN - The MPEG-4 Parametric Audio Coding Tools," in *IEEE International Symposium on Circuits and Systems*, 2000. DOI: 10.1109/ISCAS.2000.856031 4

[135] B. G. Quinn and E. J. Hannan, *The Estimation and Tracking of Frequency*, ser. Cambridge Series in Statistical and Probabilistic Mathematics. Cambridge University Press, 2001. 15, 52

[136] B. G. Quinn and P. J. Thomson, "Estimating the frequency of a periodic function," *Biometrika*, vol. 78(1), pp. 65–74, 1991. DOI: 10.2307/2336896 32

[137] R. Badeau, "Méthodes à haute résolution pour l'estimation et le suivi de sinuoïdees modulées. Application aux signaux de musique," Ph.D. dissertation, École Nationale Supérieure des Télécommunications, 2005. 85

[138] D. Rabideau, "Fast, rank adaptive subspace tracking and applications," *IEEE Trans. Signal Process.*, vol. 44, no. 9, pp. 2229–2244, Sept. 1996. DOI: 10.1109/78.536680 99

[139] L. Rabiner, "On the use of autocorrelation analysis for pitch detection," *IEEE Transactions on Acoustics, Speech and Signal Processing*, vol. 25, no. 1, pp. 24–33, Feb 1977. DOI: 10.1109/TASSP.1977.1162905 2

[140] B. Resch, M. Nilsson, A. Ekman, and W. B. Kleijn, "Estimation of the instantaneous pitch of speech," *IEEE Trans. Audio, Speech, and Language Process.*, vol. 15(3), pp. 813–822, Mar. 2007. DOI: 10.1109/TASL.2006.885242 2

[141] D. C. Rife and R. R. Boorstyn, "Single tone parameter estimation from discrete-time observations," *IEEE Trans. Inf. Theory*, vol. 20, no. 5, pp. 591–598, Sept. 1974. DOI: 10.1109/TIT.1974.1055282 19

[142] J. Rissanen, "Modeling by shortest data description," *Automatica*, vol. 14, pp. 468–478, 1978. DOI: 10.1016/0005-1098(78)90005-5 38

[143] J. Rissanen, "Stochastic complexity in statistical inquiry," *Singapore: World Scientific*, 1989. 38

[144] C. A. Rødbro, M. G. Christensen, S. H. Jensen, and S. V. Andersen, "Compressed domain packet loss concealment of sinusoidally coded speech," in *Proc. IEEE Int. Conf. Acoust., Speech, Signal Processing*, vol. 1, 2003, pp. 104–107. DOI: 10.1109/ICASSP.2003.1198727 4, 5

[145] M. Ross, H. Shaffer, A. Cohen, R. Freudberg, and H. Manley, "Average magnitude difference function pitch extractor," *IEEE Trans. Acoust., Speech, Signal Process.*, vol. 22, no. 5, pp. 353–362, Oct. 1974. DOI: 10.1109/TASSP.1974.1162598 2

[146] T. D. Rossing, *The Science of Sound*, 2nd ed. Addison-Wesley Publishing Company, 1990. 18

[147] R. Roy and T. Kailath, "ESPRIT – estimation of signal parameters via rotational invariance techniques," *IEEE Trans. Acoust., Speech, Signal Process.*, vol. 37(7), July 1989. DOI: 10.1109/29.32276 51, 81, 102

[148] B. Santhanam and P. Maragos, "Demodulation of Discrete Multicomponent AM-FM Signals using Periodic Algebraic Separation and Energy Demodulation," in *Proc. IEEE Int. Conf. Acoust., Speech, Signal Processing*, 1997. DOI: 10.1109/ICASSP.1997.599542 3

[149] B. Santhanam and P. Maragos, "Multicomponent AM-FM demodulation via periodicity-based algebraic separation and energy-based demodulation," *IEEE Trans. Commun.*, vol. 48(3), pp. 473–490, 2000. DOI: 10.1109/26.837050 3

[150] R. O. Schmidt, "Multiple emitter location and signal parameter estimation," *IEEE Trans. Antennas Propag.*, vol. 34(3), pp. 276–280, Mar. 1986. DOI: 10.1109/TAP.1986.1143830 51, 81, 83, 91

[151] G. Schwarz, "Estimating the dimension of a model," *Ann. Stat.*, vol. 6, pp. 461–464, 1978. DOI: 10.1214/aos/1176344136 38

[152] T. Söderström and P. Stoica, *System Identification*. London, UK: Prentice Hall International, 1989. 112

[153] P. Stoica, A. Jakobsson, and J. Li, "Cisiod parameter estimation in the coloured noise case: Asymptotic Cramér-Rao bound, maximum likelihood, and nonlinear least-squares," in *IEEE Trans. Signal Process.*, vol. 45(8), Aug. 1997, pp. 2048–2059. DOI: 10.1109/78.611203 15, 20, 33, 42, 50, 118

[154] P. Stoica, A. Jakobsson, and J. Li, "Matched-filterbank interpretation of some spectral estimators," *Elsevier Signal Processing*, vol. 66, no. 1, pp. 45–59, April 1998. DOI: 10.1016/S0165-1684(97)00239-9 118, 120

[155] P. Stoica, H. Li, and J. Li, "Amplitude estimation of sinusoidal signals: Survey, new results and an application," *IEEE Trans. Signal Process.*, vol. 48(2), pp. 338–352, Feb. 2000. DOI: 10.1109/78.823962 27, 28, 69, 70, 112, 114, 115, 116

[156] P. Stoica and R. Moses, *Spectral Analysis of Signals*. Pearson Prentice Hall, 2005. 27, 52, 84, 113, 115

[157] P. Stoica and A. Nehorai, "MUSIC, maximum likelihood, and Cramer-Rao bound," *IEEE Trans. Acoust., Speech, Signal Process.*, vol. 37(5), pp. 720–741, May 1989. DOI: 10.1109/29.17564 12, 84, 109

[158] P. Stoica and A. Nehorai, "MUSIC, maximum likelihood, and Cramer-Rao bound; further results and comparisons," *IEEE Trans. Acoust., Speech, Signal Process.*, vol. 38(12), pp. 2140–2150, Dec. 1990. DOI: 10.1109/29.61541 84, 109

[159] P. Stoica and Y. Selen, "Cyclic minimizers, majorization techniques, and the expectation-maximization algorithm: a refresher," *IEEE Signal Process. Mag.*, vol. 21(1), pp. 112–114, 2004. DOI: 10.1109/MSP.2004.1267055 44

[160] P. Stoica and Y. Selen, "Model-order selection: a review of information criterion rules," *IEEE Signal Process. Mag.*, vol. 21(4), pp. 36–47, July 2004. DOI: 10.1109/MSP.2004.1311138 38, 39, 40, 41, 109

[161] P. Stoica and K. Sharman, "Maximum likelihood methods for direction-of-arrival estimation," *IEEE Transactions on Acoustics, Speech and Signal Processing*, vol. 38, no. 7, pp. 1132–1143, 1990. DOI: 10.1109/29.57542 51

[162] P. Stoica and K. Sharman, "Novel eigenanalysis method for direction estimation," *IEE Proceedings F Radar and Signal Processing*, vol. 137, no. 1, pp. 19–26, 1990. 51

[163] P. Stoica and T. Söderström, "On reparameterization of loss functions used in estimation and the invariance principle," *Elsevier Signal Processing*, vol. 17, pp. 383–387, 1989. 49

[164] P. Stoica, Z. Wang, and J. Li, "Robust Capon beamforming," *IEEE Signal Process. Lett.*, vol. 10(6), pp. 172–175, June 2003. DOI: 10.1109/LSP.2003.811637 101

[165] P. Strobach, "Low-rank adaptive filters," *IEEE Trans. Signal Process.*, vol. 44(12), pp. 2932–2947, Dec. 1996. DOI: 10.1109/78.553469 98, 99

[166] A. Swindlehurst and P. Stoica, "Maximum likelihood methods in radar array signal processing," *Proc. IEEE*, vol. 86, no. 2, pp. 421–441, 1998. DOI: 10.1109/5.659495 49

[167] D. Talkin, "A robust algorithm for pitch tracking (RAPT)," in *Speech Coding and Synthesis*, W. B. Kleijn and K. K. Paliwal, Eds., chapter 5. Elsevier Science B.V., 1995. DOI: 10.1109/89.326635 2

[168] J. Tabrikian, S. Dubnov, and Y. Dickalov, "Maximum a posteriori probability pitch tracking in noisy environments using harmonic model," *IEEE Trans. Audio, Speech, and Language Process.*, vol. 12(1), pp. 76–87, Jan. 2004. DOI: 10.1109/TSA.2003.81995010.1109 2, 32

[169] T. Tolonen and M. Karjalainen, "A computationally efficient multipitch analysis model," *IEEE Trans. Speech Audio Process.*, vol. 8, no. 6, pp. 708–716, Nov. 2000. DOI: 10.1109/89.876309 2

[170] G. Tzanetakis and P. Cook, "Musical genre classification of audio signals," *IEEE Trans. Speech Audio Process.*, vol. 10(5), pp. 293–302, July 2002. DOI: 10.1109/TSA.2002.800560 6

[171] S. van de Par, A. Kohlrausch, R. Heusdens, J. Jensen, and S. H. Jensen, "A perceptual model for sinusoidal audio coding based on spectral integration," *EURASIP J. on Applied Signal Processing*, vol. 9, pp. 1292–1304, 2004. DOI: 10.1155/ASP.2005.1292 3

[172] S. van de Par, J. Koppens, A. Kohlrausch, and W. Oomen, "A new perceptual model for audio coding based on spectro-temporal masking," in *Proc. 124th AES Convention*, 2008. 3

[173] A.-J. van der Veen, E. F. Deprettere, and A. L. Swindlehurst, "Subspace-based signal analysis using singular value decompostion," *Proc. IEEE*, vol. 81(9), pp. 1277–1308, Sept. 1993. DOI: 10.1109/5.237536 81

[174] P. van Overschee and B. D. Moor, *Subspace Identification for Linear Systems: Theory, Implementation, Applications*. Kluwer Academic Publishers, 1996. 81

[175] E. Vincent, N. Bertin, and R. Badeau, "Harmonic and inharmonic nonnegative matrix factorization for polyphonic pitch transcription," in *Proc. IEEE Int. Conf. Acoust., Speech, Signal Processing*, 2008, pp. 109–112. DOI: 10.1109/ICASSP.2008.4517558 2

[176] D. Wang and G. J. Brown, Eds., *Computational Auditory Scene Analysis: Principles, Algorithms, and Applications*. IEEE Press, 2006. 4

[177] M. Wax and T. Kailath, "Detection of the number of signals by information theoretic criterion," *IEEE Trans. Acoust., Speech, Signal Process.*, vol. 33(2), pp. 387–392, Apr. 1985. DOI: 10.1109/TASSP.1985.1164557 87, 88

[178] W. Xu and M. Kaveh, "Analysis of the performance and sensitivity of eigendecomposition-based detectors," *IEEE Trans. Signal Process.*, vol. 43, no. 6, pp. 1413–1426, 1995. DOI: 10.1109/78.388854 88

[179] B. Yang, "Projection approximation subspace tracking," *IEEE Trans. Signal Process.*, vol. 41(1), pp. 95–107, Jan. 1995. DOI: 10.1109/78.365290 99

[180] M.-Y. Zou, C. Zhenming, and R. Unbehauen, "Separation of periodic signals by using an algebraic method," in *Proc. IEEE Int. Symp. Circuits and Systems*, vol. 5, 1991, pp. 2427–2430. DOI: 10.1109/ISCAS.1991.176066 3

[181] E. Zwicker and H. Fastl, *Psychoacoustics - Facts and Models*, 2nd ed. Springer, 1999. 3

About the Authors

MADS GRÆSBØLL CHRISTENSEN

Mads Græsbøll Christensen was born in Copenhagen, Denmark in March 1977. He received the M.Sc. and Ph.D. degrees in 2002 and 2005, respectively, from Aalborg University in Denmark, where he is also currently employed at the Department of Electronic Systems as Assistant Professor. He has been a visiting researcher at Philips Research Labs, Ecole Nationale Supérieure des Télécommunications, Columbia University, and University of California Santa Barbara. Dr. Christensen has received several awards, namely an ICASSP Student Paper Award, the Spar Nord Foundation's Research Prize for his Ph.D. thesis, and a Danish Independent Research Council's Young Researcher's Award. His research interests include digital signal processing theory and methods with application to speech and audio, in particular parametric analysis, modeling, and coding.

ANDREAS JAKOBSSON

Andreas Jakobsson received his M.Sc. from Lund Institute of Technology and his Ph.D. in Signal Processing from Uppsala University in 1993 and 2000, respectively. Since, he has held positions with Global IP Sound AB, the Swedish Royal Institute of Technology, King's College London, and Karlstad University. He has also been a visiting researcher at King's College London, Brigham Young University, Stanford University, Katholieke Universiteit Leuven, and University of California, San Diego. He is currently Professor of Mathematical Statistics at Lund University, Sweden. He also holds an Honorary Research Fellowship at Cardiff University, UK. He is a Senior Member of IEEE, a member of the IEEE Sensor Array and Multichannel (SAM) Signal Processing Technical Committee, and an Associate Editor for the IEEE Transactions on Signal Processing, the IEEE Signal Processing Letters and the Research Letters in Signal Processing. His research interests include statistical and array signal processing, detection and estimation theory, and related application in remote sensing, telecommunication and biomedicine.

Printed in the United States
by Baker & Taylor Publisher Services